Telling the Bees

TELLING THE

BEES

An Interspecies Monologue

Dominic Pettman

Fordham University Press   New York   2025

Fordham University Press has no responsibility for the
persistence or accuracy of URLs for external or third-
party Internet websites referred to in this publication
and does not guarantee that any content on such
websites is, or will remain, accurate or appropriate.

Fordham University Press also publishes its books in a
variety of electronic formats. Some content that appears
in print may not be available in electronic books.

Visit us online at www.fordhampress.com.

Library of Congress Cataloging-in-Publication Data
available online at https://catalog.loc.gov.

Printed in the United States of America

27 26 25   5 4 3 2 1

First edition

Preface

"Aren't You Ashamed?"

I still vividly recall trying to explain the interactive mechanics of Facebook, in the early days of what is now called social media, to my bemused mother—a woman who frequently claims to have outlived her own century. I explained how I could now post photos of family gatherings, or nights on the town, or simply my lunch, on any given day. I detailed how—at the click of a button—I could share a joke, or a book recommendation, or a stray thought, along with more important news like a change of career or living situation. Further, I noted that people electronically connected to me could then respond with "likes" or smiley faces, or thoughts of their own. My mother blinked at me for a moment, processing the very notion of this new kind of pseudo-communication, enabled by that nascent omni-technology called the Internet. "But," she ventured, bestowing upon me her most earnest expression, "aren't you *ashamed*?"

At the time, I laughed off her disapproving perplexity. In hindsight, however, I can now confidently reply: *Of course* I am ashamed. Just as we are *all* ashamed—or at least should be—when we willingly and compulsively contribute to these predatory corporate platforms that hijack our profoundly human desire to interact with one another. In providing an isolating parody of "actually existing socializing," these platforms deliberately invite us to be satisfied with a menu, rather than a meal. Admittedly, we've quickly learned to metabolize this shame, so we don't feel its sting most of the time (even as countless studies have shown that prolonged use of social media is often followed by enhanced feelings of depression, loneliness, envy, and other negative emotions). Indeed, we're all familiar with the existential deficit that occurs when we post on Facebook or Instagram, or what most people still tend to call Twitter. Every time we "put ourselves out there"—for all our friends and acquaintances to see, and ideally to re-acknowledge our existence—we are usually rewarded by only a handful of paltry responses, if anything at all. Our visibility and "reach" depend on the opaque whims of the algorithm, which seemingly has anything but our own interest at heart. (Were the algorithm able to speak—and speak freely—it would probably admit, "I'm just not that into you.") And yet we continue to post under the sway of a specific kind of mediated "cruel optimism" in the hope that we may, finally, be validated, appreciated . . . *seen*. Of course, this becomes a vicious cycle, encouraging us to post *even more*, in a vain attempt to mitigate the emotional deficit through a constant operation of diminishing returns and increasingly hollow non-encounters.

In my case, the shame is compounded by the fact that I have published a book-length critique of the malign machinations of Zuckerberg and his spawn, and I regularly teach a college course called "Anti-Social Media: Addiction, Distraction, Attention." Surely I am therefore a Zen master in online restraint and digital hygiene. As any of my online connections could tell you, however, rarely a day goes by that I don't succumb to the temptation to add my voice to the global typographic cacophony. (Proving yet again that being *cognitively aware* of the fact that something is harmful does not necessarily serve as a prophylactic against continuing to consort with that very same something.) Increasingly, however, the mental dissonance of complaining about social media—while still tossing my daily *bons mots* into the roiling digital void—began to weigh on me. And so, in the autumn of 2019, as the long, warm days began to cool and shrink, I decided to experiment with a different form of communication, one untarnished by the insidious solicitations of Silicon Valley. One supported by centuries of tradition. Namely, *telling the bees*.

The Tradition of "Telling the Bees"

I first heard about this venerable custom in one of my undergraduate history classes on European folk culture. We learned how any news worth sharing throughout the village, beyond the most immediate neighbor, was sometimes also considered to be intelligence worth informing the local beehives about, a practice underwritten by the more intimate relationships we enjoyed back then, in more bucolic times, with our fellow creatures. Funerals were the most common form of telling the bees. Sometimes the hives themselves were draped in black material or tied with black ribbons, while others were moved to a different position, to respect the deceased, and in honor of a fresh start. In the unfortunate case of the death of a beekeeper, the next in line (usually a brother or a son) would knock on the hive walls and announce the transition. This was considered the most respectful way of ensuring that the new master would be respected. Indeed, it was felt that if the local apiaries weren't properly "put into mourning"—in sympathy with the nearby human abodes—then the bees might refuse to produce honey, or simply up and leave. Such abandonment might well augur more general bad luck to soon befall the community. While the superstitious logic here is somewhat obscure, the bees' potential refusal to do their apian duty is perhaps connected to a more-than-human sense of possible resentment at being kept "out of the loop" of local affairs. (As the Elizabethan playwright John Lyly put it, "No creature is more wreakful, nor more fervent to take wreak, than is the Bee, when he is wroth!") Grief, at any rate, was powerful enough to cross the species divide—at least from the human side of reckoning. And while today we may consider this a flagrant case of anthropomorphism, a habit of bee-telling also speaks of a more inclusive approach to other life-forms, no matter how diminutive or alien.

The advent of bad news, however, wasn't the only occasion with which to motivate the locals to send a messenger out into the fields. Births, engagements, weddings, windfalls, and new arrivals in town were all opportunities to wander down to the colony and spread the happy word. Sometimes leftover cake and wine from recent celebrations were left inside the hive, to ensure that the apians felt included in the festivities. If, on the other hand, a beekeeper happened to drink too much mead—that heady concoction his own buzzing charges helped create—he may well feel obliged to apologize the next morning, especially for any linguistic liberties taken. (This because bees were also known to be impatient with cursing and other forms of bad language.)

The general practice of telling the bees—well known in the British Isles, and recorded across western continental Europe, before being brought to New England, and beyond; even noted among some peoples of East Africa—has its roots in Classical Greece. (Historians are divided, however, on the question of whether this is the origin of the hippie phrase, most famously chanted in *Jesus Christ Superstar*, "What's the buzz? / tell me what's a-happenin'.") Certainly, there has been a persistent sense in which bees *have*, or at least dimly represent, "souls"—in their industry, collective social wisdom, their seemingly divine gift of golden honey, as well as their capacity to ascend toward the heavens. (In ancient Egypt, bees were considered to be reincarnated humans, while some surviving sources describe them as "winged messengers of the gods.") It makes perfect sense, then—as an especially diligent instance of God's creation—that bees have traditionally been inserted into our own local media networks, almost as honorary persons. ("Telling the bees" being one pragmatic way to further reflect on the latest development in God's great plan.)

Remarkably, this custom continues today and even made the headlines recently, when Queen Elizabeth II passed away. A no doubt capable fellow by the name of John Chapple—who also happens to be the Royal Beekeeper—followed folkloric protocol by moving from hive to hive, announcing, "The mistress is dead." He then turned on his heel and retraced his steps, adding in a loud, clear voice, "But don't you go. Your master will be a good master to you." (The master being, of course, King Charles III.)

Interspecies Monologues and Asocial Media

From the perspective of our modern age, it may seem presumptuous to assume that insects—no matter how complex and organized—will be interested in human doings. As I have already intimated, however, there is something charming about the desire to share our latest news beyond our own kind. Indeed, I like to think of this urge as an admirable inclusive gesture, one based on a respect for the intelligence and curiosity of the bees and, by extension, the natural world more generally. Moreover, I suspect that the majority of those bygone human souls—who opened the back garden gate and wandered down to the fragrant combs under the blossoming apple trees—would sincerely be interested in any news the bees decided to tell them in return, if only they themselves understood the buzzing language of the apiary.

Indeed, it was in this spirit that I decided to embark on my latest conceit: sharing my thoughts with some proverbial bees, buzzing in my own head, rather than with the swarming, waspish, "hive mind" of the Internet. ("These bees," asked a colleague, playfully, when I told her of this project. "Are they in the room with us now?") Given my job as a professor of media and culture, I am obliged to admit upfront that I could not fully resist the siren song of social media, because it is now the main portal to "the discourse," which is itself the bread-and-butter of my business. These networks, of course, have their benefits and seductions, which is why they can be so hard to resist. (Lord knows, I would have trouble connecting with my students at all if I didn't have at least some passing familiarity with recent memes, slang, or scandals.) By the same token, in embarking on this new, more reticent mode, I found it a relief to no longer be chasing "likes" or "hearts"—or even substantive responses—from my fellow humans online. (A group that I had already whittled down on Facebook to something approximating the famous Dunbar number of 150, above which point any "organic" community is said to fragment into multiple degrees of separation.) Indeed, I found immediate comfort in talking to my new bee friends—imaginary avatars based on some of the actual bees I encountered on my daily walk through Central Park. Unlike my "friends" on social media, these were under no obligation to respond. As the initial possibility of a Trump presidency loomed, I had an outlet that would not devolve quickly into pedantry and competitive kvetching, according to the universal human law of "the narcissism of minor differences." And when the word *coronavirus* first seeped into the headlines, I had access to a demographic that I could rely on not to splinter into conspiracy theorists, on the one hand, or shrill alarmists, on the other.

So to say, in the age of "cottagecore"—an aestheticized yearning to leave our devices at home, put on hand-woven garments, and go outside to "touch grass"—I took a new, natural leaf from the zeitgeist and tried to get in touch with my roots. Of course, in an age in which honey may turn blue, because of chemicals in the water—and where the number of actual bees has fallen vertiginously each season—this could not simply be a Thoreau-like exercise in agrarian cosplay. Inevitably, my hypothetical bees were obliged to hear much about today's technologically inflected (and infected) environment, to the degree that it makes little sense to think of a pristine "nature" somehow untouched by our own human doings, in this perhaps terminal stage of the Anthropocene.

Undoubtedly, I found great comfort in addressing a very different kind of creature with my daily thought processes, now unburdened by the kind of anticipation and second-guessing that goes into any social media post. At the beginning of the experiment, I sought out real live bees to talk to (or perhaps *at*). These tended to be buzzing around bushes near Shakespeare Garden, covered in pollen; sometimes fast asleep inside a flower, their fuzzy butts sticking up in the air. As the project evolved—and as we humans were driven inside during the pandemic—I was obliged to speak to the memory of such bees, speculating on their current welfare and whereabouts. (Sadly, my co-op board would never approve a beehive on the roof of the building, as I'd initially hoped.) The actuality of the addressee, when it comes to literature especially, is of course a slippery phenomenon. Whom indeed are we writing to, when we write a love letter? Is it the current beloved? Or one's ex, in disguise? Or our future paramour, in fact? Or simply ourselves? (Roland Barthes is insistent that we are always writing for *someone*, even as that someone can shimmer and change, while Mark Fisher feared that any love letter addressed to ourselves is always already mis-addressed.) The reality of my bees was always hovering at the very edges of my perception. Nevertheless, they were my constant (absent) companions.

The pages that follow are not at all "academic" or "scholarly," even as they may allude to certain ideas or experiences originating from that part of my brain or vocation. Rather, they can be read as an intimate journal of the last few years, from the vantage point of a transplant to New York City: a series of years in which the world truly slipped off its already wobbly axis: politically, ideologically, economically, culturally, environmentally, and so on. Alternatively, this book can be read as an epistolary memoir in which the addressee is not an estranged friend, or distant lover, but a virtual animal totem: a specific figure, at once real and mythic, through which I try to make personal sense of a world in flames. The bees buzzing around my own head helped me deal with the violent vibrations of hyper-entropic capitalism, just as a white noise fan can help screen out industrial noises. They kept me company as I attempt-ed to escape the feverish mediascape, while also being obliged—and unable—to tear my horrified gaze from it, forever trying to make sense of the spectacle.

A skeptic may say that it is a foolhardy experiment, to spend so much time pretending to communicate with a "lower" life-form, untutored in the nuances of the English language. A *different stripe* of skeptic, however, may say precisely the same thing, about those who reg-ularly post their thoughts to the troll-ridden, bot-filled hellscape of social media.

Telling the Bees

October 19, 2019

My uncle was a beekeeper. He kept bees. He kept them in boxy hives that reminded me of smaller versions of the houses I had seen in tropical climates, striding immovably above the potentially flooded ground on wooden stilts. I never saw my uncle actually wearing those strange, billowy beekeeping outfits. I did, however, see one hanging on the wall once, looking to my boyhood mind like a deflated spacesuit. I somehow associated it with the hazmat emergency suits I saw in the disaster movies of my youth. Indeed, my father told me that one day, several years before, as my uncle was moving a hive from one place to another, the bottom of the box fell away, and hundreds of confused and angry bees fell into the open tops of his boots, stinging his lower legs and feet with such vigor that my poor uncle couldn't walk for weeks. This scenario was appalling to me, and I took every opportunity to avoid visiting the bees thereafter. (Even though I was assured by my uncle that they could be made drowsy with smoke.)

This is probably why I'm a little tentative, here at the beginning of our relationship. Since, yes—I'm a bit skittish around you bees, with this image of an almost mythical mass stinging event I have carried with me since I was very young. It's true, however, that I managed to survive a rather classic 1970s free-range childhood in which I would walk about in bare feet across clover all the time, with only a couple of stings. (Thankfully I don't seem to be allergic to your kind. Had that been the case, I would have been obliged to tell you things—to bring you the news of, and from, the humans—from afar, by proxy.) As it is, I'm glad I have this opportunity now: to come to you each day—where I happen to find you, busy among the bushes—and start my own personal version of the old custom of "telling the bees." Indeed, this time-honored activity—practiced in villages all over Europe, for centuries—seems much healthier to me than confessing things to the digital ether, to the anonymous world via social media (which is perhaps a perverse extension of this tradition). Unlike my online friends, you bees will never "like" what I say, nor comment on its content, which will of course be a relief. It will be a relief, as I expect nothing from you, nothing but a waxy, distracted kind of listening. (Purely imagined, on my part.) So to say, the sound you bees make together is the only kind of "buzz" I care about at present. You are the only "hive mind" with which I want to spend time.

Of course, I hope you find what I have to say interesting, on some level. No doubt your ancestors have heard some wonderfully scandalous gossip and luridly fascinating stories, thanks to generations of people wandering down the garden path at dawn to keep you abreast of the comings and goings of the village. At present I cannot pretend to have anything especially compelling to relate. Nevertheless, I am sure we will get to know one another through this process.

In the meantime—for the first few visits at least—I hope you are not offended that I tuck my pants into my shoes . . . just in case.

October 20, 2019

Unlike my uncle, I am not a beekeeper. I do not keep bees.

Now that I have an irresistible urge to tell the bees so many things—trivial things, important things—I suppose I should explore whether it's possible to keep a hive or two, given my circumstances (living in the middle of a big city). I could possibly become one of those urban beekeepers I've read about. But the logistics seem daunting. Moreover, I would have to get permission from the co-op board, and then from the management company, and perhaps even from the city. No doubt one of the bovine children of my neighbors is allergic to bees, so the idea would be dismissed before I could even make a good case. (The hives would have to be kept on a shared roof deck.) So I will have to content myself with telling the bees that I find on my morning walks through Central Park, and in the Shakespeare Garden. (Or, when it's raining, the virtual bees that buzz around my mind.) I like to think that whichever bee I happen to chance upon, it will pass along my message to some kind of vast hive network, so all the others in the greater vicinity will be kept abreast of my thoughts; and so the next random bee I encounter will already be familiar with my recent reports. (Insectoid games of telephone notwithstanding.)

Tomorrow I will go in search of bees in the Park, to tell of my dreams, my hopes, my fears—my whimsical daydreams. Perhaps I will find two curled up together, asleep in a flower, holding each other's legs, and covered in pollen. These I will not disturb and will instead move on to one of their more alert siblings. (I recently read that bees can leave their homes extra early, to secure access to the most delicious flowers, only to fall asleep on the bloom until such time as they feel energized enough to begin their daily labors.) Indeed, we shall have to talk about this soon, you bees and I. The question of toil. I like to think of nature as free of the drudgery of work. But your drone-like obligations make me pause. Does your labor feel like a joy to you? Do you love your job? Or do you, like me, resent the need to head out into the world at dawn in order to make a living?

October 22, 2019

Good morning, bees. I hope this day finds you in good spirits, despite the challenges that refuse to evaporate with each new morning. (Does that sound too much like an e-mail from HR? You'll have to forgive me if it takes me a while to find the right tone. I've never had an extended conversation with a group of non-humans before.)

The sun is about to rise on an unseasonably warm autumn day. Few can deny climate change anymore, as we city dwellers hustle past store windows featuring warm, fleecy coats while still wearing t-shirts and shorts. I recently read that you bees were voted "the most important creature in the world" by a group of esteemed scientists. Congratulations. There is, however, a sting in the tail, isn't there? This is more a consolation prize, given the fact that only 10 percent of the world's bee population remains, according to the same scientists. I wish I knew what to say about this, other than "I'm sorry" or "This is terrible." Birds are vanishing. Forests are falling silent. It's bleak. It's unthinkable. But we humans carry on, destroying the environment through our greed, laziness, and apathy.

No doubt you know this. You don't need another human trying to unburden his guilt. You're not my therapist, or my priest. I don't expect any wisdom or absolution from speaking with you. That would be a final insult.

Once upon a time—not that long ago, really—we humans told the bees of happy events: weddings, birthdays, anniversaries, new members of the community, and such. I shall try to continue this tradition. But it is difficult when bees are harder and harder to find, because of the human blight. All this has been weighing on my mind because the media—a kind of electronic version of your waggle dance, I suppose—have been telling us about a young girl from Sweden. This young girl sailed a yacht all the way across the Atlantic to New York to tell the human leaders what a terrible job they have been doing, as elected stewards of the Earth and shepherds of the future. I don't mind telling you that I misted up as I watched this girl expressing the anger we all feel—at the rich, and at ourselves—for letting all the other creatures and living things down. My face twisted in sympathy with hers, in a last-ditch rictus plea to those in power to change their ways.

The so-called adults in the room did not respond well. They do not like to be reminded of the possibility of shame, as so much of their activity depends on denying that such a thing has taken root in their arid souls. The great nausea of *true* shame, as opposed to fleeting embarrassment, comes from a sense of being suddenly reminded that we are alive, that life isn't really a game (or if it is, it is a game with high stakes), and that we cannot escape the burden of being conscious beings as long as our confused hearts keep beating. Shame goes beyond psychology into metaphysics, because it reminds us of the predicament of awareness-within-existence. And as such, it confronts us with the fact that we almost always squander the gift of the present. Humans are experts at denial, you see, and other forms of repression.

A famous German philosopher once claimed that the modern world is founded on a kind of willed, collective amnesia concerning the miracle of existence. We behave as if our precious, flickering lives on this lonely, beautiful planet are a given, a banality. Something to take in one's stride, without a second thought, while pursuing plastic trinkets and shadows.

I tend to agree.

In denying the magic of life, we have removed the conditions of its flourishing.

Our great shame: "The forgetting of bee-ing."

October 24, 2019

I saw a documentary recently in which some of your bee cousins made a hive out of plastic. The film seemed to suggest that this was possibly an improvement—if we can put aside our biases toward tradition—because the plastic kept parasites away. But I can't imagine that bees raised in this inorganic environment would be very happy. It reminds me of the blue honey I read about, created by bees that lived near a factory that produced M&M's. Weird ecology.

In the evening, at twilight, I look out at giant buildings, illuminated from within, like human hives. Except we are busy producing money, rather than honey. (And mostly for someone else.)

The ancient Greeks encouraged us to extract the "sweetness" that can be found in the comb of life. Today, with our more sophisticated palates, we pursue *umami*. This in turn makes me wonder how the world might be different if Japanese philosophers had been as influential as the Greeks, in a world-historical sense. The sense of taste is surely a decisive aspect of our thought patterns.

What do you think about, my precious bees, as you sample your serendipitous *omakase*, from petal to petal?

October 25, 2019

Things are changing. And not for the better.

Today I saw a presumably homeless person in a large fountain, in the middle of the Park. He was shirtless and stuffing the pockets of his soggy pants with coins that tourists had thrown into the water along with a wish. This seemed to be the breaking of an honor code, or a taboo: an act of special desperation. Times are becoming so harsh that even other people's hopes are no longer worth respecting. Together, these silver wishes may add up, here in the present, to a stale bagel or some watery soup.

On the flip side, I confess to misguided daydreams, the whimsical kind bred from comfort and privilege. In these daydreams, academics like me roam the land like displaced sharecroppers in search of work. Instead of trudging to the same ugly building to teach the same underwhelming classes to the same impassive faces, professors would follow the pedagogical harvest according to weather patterns, the suggestive slope of valleys, and the rumors of possible work one overhears in a splintered alehouse, or around a chili pot bubbling over an open fire. ("Word on the corner is that some folks up near Jacksonville are hankering for some knowledge about counter-narratives in lesbian modernist aesthetics.") We would warm our cracked hands over flaming oil drums and share syllabus tips before jumping onto a boxcar heading south, in search of fresh fields of bobbing heads in need of cultivation, and in the hope of a paycheck with which to mop the sweat on our brows. And during the full harvest moon, on a warm early autumn night, we would dance the somewhat obscene Dance of Best Practices to the sound of the jug and the fiddle, flirting with the lower-level administrators who forever sit outside the circle of shuffling nomadic educators, keeping a watchful eye on our labors. There they sit, on hay bales or half-broken office chairs, spitting chewing tobacco into the dirt, with a wry grin, at the presumption of our tipsy overtures, made in an arcane language, seductive only to the uselessly anointed.

October 29, 2019

After a seemingly interminable dry spell, it finally rained yesterday. I'm sure you noticed the soft pitter-pattering on the waxy roof of your hive, wherever that is, for most of the day. Walking around the Park this morning was an olfactory symphony. The half-dead leaves were still soaked from the drizzle and released various fragrances—from over-brewed tea to damp tobacco. Some sections smelled downright hallucinogenic. I could imagine little elf-like creatures—the ones my eyes are never fast enough to catch, but that I can sense blurring around me, now and again—harvesting the damp leaves, drying them out in front of their little potbelly stoves, and then smoking them in long-stem pipes fashioned out of twigs and acorn shells. Perhaps inhaling such powerful vegetation gives the elves' pointy ears such sensitivity that they can still hear the faintest of echoes from the famous Simon & Garfunkel concert in the Park here, in 1981. ("They've all come to look for Ameeeeeericaaaaaa.")

In any case, I hope the remaining pollen has fluffed up again for you. Perhaps it's a treat now, like yellow marshmallow.

After my morning constitutional, I started reading for a new class I'm teaching: "What Was the Human?" This week's thought-food was written by a man who is still convinced of the absolute unique nature of my species (which he automatically assumes embodies a type of natural superiority). Very jarring, to hear such humanist chauvinism in 2019, after reading so many recent critiques of anthropocentrism. But in some ways, this rather arrogant fellow does have a point. Humans *are* unique. But as I have said before, human exceptionalism is much like American exceptionalism. The latter may be an objective fact, in certain ways, but that doesn't mean it's something to be proud of, or to try to emulate.

Despite such pompous voices on the syllabus, I'm enjoying the course because it helps me think about the so-called human–animal divide, which I feel can be minimized, or at least partly bridged, through exercises like this one: "telling you, the bees."

Indeed, they say that animals are good to think with.

But they are even better to talk with.

November 3, 2019

Good afternoon, apians. A late start today, writing my little bee-stiary.

I realize I have not been good at keeping you abreast of the news, as is customary, when it comes to this particular ritual in which we're both engaged. (Or, at least, in which I attempt to engage you.) Normally I would give you significant updates of the goings-on in the village, but the village I live in is vast. I suppose I could give you the gossip about my building, which houses maybe 150 souls, but in truth I know next to nothing about them.

Meanwhile, the larger news involves the probable impeachment proceedings against the orange Idiot King, who currently spreads his foolish reign like an apocalyptic toxic Cheez Doodle accident across this ravaged land. Not having grown up in this country, I find the actual mechanics of impeachment vague. But I like to imagine them as a kind of reversal of the Japanese folk tale I loved as a child, in which a young boy is born from inside a giant peach. I picture instead a stiff and formal "in-peachment" process, leading inexorably to the confinement of the sitting president inside a giant fruit, bioengineered expressly for this purpose, and boasting similar coloring to the disgraced statesman's own sandpapery skin. The Idiot King would be stripped of his executive crown, live on television, and stuffed—both ceremoniously and not—into the middle of the pulpy peach, where the stone should be, but where now his stony heart, and the grotesque fleshy husk that has hitherto housed it, has been wedged and compressed. The nation would cheer, as that hideous mouth, which could never stop flapping, was finally muffled, and then silenced forever, by the airless fibers and juices streaming into his dimming consciousness.

Smiles would suddenly return to the streets of this city, and strangers would trade them like precious tokens. Smiles prompted by successful in-peachment proceedings in the Capitol.

November 4, 2019

I was a little delirious yesterday, wasn't I, dear bees?

Perhaps because I haven't been sleeping especially well, in part because of the exhaust fans just outside my bedroom window, which have been very much the sonic thorn in my side since I've moved to this part of town. There is also a mysterious nocturnal internal hum that can test my sanity, obliging me to drag my mattress into the little hallway linking my front door to the kitchen, where the resonance is somewhat diminished, though not enough to give me peace.

This makes me wonder what it is like to sleep in your own combs. Does the buzzing die down at night? Or do you bees also hum after sundown, providing a kind of cozy, golden noise machine in which your insectoid bodies like to slumber? Or perhaps some of you are driven mad by the incessant din and choose instead to fly away at night, toward the river, in search of a lonely, buzzless perch on which to rest your weary wings.

I also wonder whether someone has taken the time to map out a taxonomy of humming noises, separating the benign from the malefic. On the pleasant end of the spectrum would be the hums of young maidens, daydreaming of amorous adventures, as well as the hum of your own hive-like activities in an English garden on a sleepy summer afternoon. On the other end of the scale are the electronic hums of the machines on which we have, in our helplessness, come to depend. Massive refrigerators filled with semi-edible chemicals, compressed into the rough shapes of former foods. Ice machines in mid-priced hotels, quarantined into dark and dingy rooms at the end of the corridor, vomiting ice into a bacteria-filled overflow tray at irregular intervals, all day and all night, like some kind of wizard's punishment for a frat boy who went too far and was now forced to remain in a mechanical, drunken stupor for all eternity.

I read an article recently which explained that gorillas and chimps have been known to hum special food songs while they are eating. This is, of course, adorable. And a far cry from the industrial hums that haunt us daily, subliminally or not, pressing against our eardrums like aural acupuncture needles. The brain-buzzing of engines, of fans, of generators. Of appliances applying themselves. And of unconditioned conditioners and rusting mechanisms, designed to keep us held aloft in the transparent suspense of Perspexual perplexity.

If the coming social collapse will rid us of this ambient clamor, then I shall consider that a silver lining, so at least the animals can slumber more peacefully, amidst the new muteness of the machines.

November 11, 2019

Today, in your honor, I taught a class on Ernst Jünger's strange and prescient novel *The Glass Bees*. The students seemed to appreciate it as much as I did, which was a relief, because very little actually happens in the story. The entire novel is composed of the thoughts of an ex–military man, waiting in a luxuriant garden for an Elon Musk–type figure to arrive and grant him a job interview for an industrial liaison position. This man's name is Zapparoni: an entrepreneur and scientific genius who became rich by designing new technologies for the military-industrial-entertainment complex, everything from tiny nanotech turtles ("It's turtles all the way down") to artificial movie stars, indistinguishable from the real thing. While the ex–military man waits, we are given access to his thoughts about the passing of time, and the changing of epochs, thanks to this new wave of technologies. ("How was it possible that the times darkened so quickly—more quickly than the brief span of a lifetime, of a single generation?") The climax of the novel—if we can call it that, for such a low-key moment—is when the job applicant notices that the bees that have been buzzing around him are in fact made of glass. Captain Richard (for this is the applicant's name) reels at this realization, in a moment of existential disorientation and negative epiphany.

After recovering from his initial shock, Captain Richard observes these artificial bees carefully. (Jünger was an entomologist, as well as a soldier and writer.) "I at once had the impression of something undreamed-of, something extremely bizarre—the impression, let us say, of an insect from the moon. A demiurge from a distant realm, who had once heard of bees, might have created it." Having paid some more attention to these tiny, industrious flying machines, Captain Richard decides that Zapparoni's creatures proceed "more economically" than real bees—that is to say, "They drained the flower more thoroughly." The captain is disturbed by the cold efficiency he is witnessing, where "the vital force of the flowers" seems to be exhausted, after being touched by the small drones' glass probes. Turning to the nearby home of these glass bees, the narrator notes that this resembles "less a hive than an automatic telephone exchange." Rather than organic, waxy openings for the bees to come and go, "the entrances functioned rather like the apertures in a slot machine or the holes in a switchboard." The captain is perturbed, not only because the bees are man-made but also because they seem to be obeying a ruthlessly efficient program, unknown in the waverings and meanderings of nature. (No bees sleeping on the job here!) . . . "It was evident," he thinks, "that the natural procedure had been simplified, cut short, and standardized." Indeed, "the whole establishment radiated a flawless but entirely unerotic perfection."

Isn't that just the perfect description of humanity's hypnotic state, in relation to our ever-evolving technologies? *Unerotic perfection*. For some perverse reason, this seems to be the endgame of so much man-made activity, so much prattle and hassle and rattling buzz.

In any case, I didn't mean to give you a sermon today, dear bees. Or impromptu moral instruction. I'm sure you're aware that several labs and universities around the world are trying to make Jünger's queasy nightmare a reality, now that we seem to have nearly wiped our humble sponsors off the face of the Earth. If I could send you all to sting the people who authorize such projects, I would. But your erotic imperfection would no doubt ensure that you would become distracted from this mission, as you became seduced by a flower, or a nap (or a nap on a flower) before reaching your target.

For this I love you.

And for this, I worry about you so.

November 12, 2019

Have I already mentioned that I'm a teacher? Is there a form of pedagogy to be found in the apiary? Is your honey combed with the sweet glow of insectoid knowledge? Moreover, does the hive mind of the bee grow organically, or are there mini-lectures contained in the waggle dance that humans are too crude and too slow to perceive a kind of intelligent choreography that transmits a universe of wisdom, so much more than the location of new sources of food (the only hypothesis our dull scientific minds can come up with as a rationale for that same dance). Perhaps the waggle contains multitudes: a soft swarm of subtle absorptions of the nature of things—part koan, part treatise. Perhaps worker bees benefit from royally subsidized education programs, studying at night in order to earn special status or standing.

Certainly, my students are still a long way from the getting of wisdom. Not that I was a model student at their age; the hormones rushing in my ears like the white-water torrent of an underground river, heard between two compressed sheets of limestone. And today attention to one's studies is even more challenging, now that we all have a glowing fondle-slab in our pockets, whispering wordless prospects that lurk *just behind* whichever app you happen to have installed, vibrating beneath itchy fingers. Fantastic haptics.

November 14, 2019

Imagine the kind of world, dear bees, we would be living in if, instead of a staid and gentle Shakespeare Garden in Central Park, there was a Rabelais Rockery.

November 15, 2019

Good evening, dear bees.

In an effort to get better acquainted with your ways, I borrowed a classic book in the canon of apian lore—Maurice Maeterlinck's *The Life of the Bee*, first published in 1901. A more recent treatise, with much more up-to-date and scientifically verified information, would have been the wiser choice, perhaps. But this lengthy essay has a celebrated lyricism that I thought would lend a more sympathetic tone to our new relationship. After all, Maeterlinck is—or at least was—well known for displaying what we would today call a "more than human" appreciation for your kind, refreshingly unencumbered by the species exceptionalism that we find in many scientists even today. Indeed, this erudite amateur beekeeper is already proving to be a beguiling guide to the hive, even as his language can be overly florid, in that classic maximalist late Victorian style.

From the opening pages, I am struck by how respectful and admiring the author is in relation to his subject. He does not have the positivist scientist's grasping drive to *know*, even as he exhibits a healthy and lasting curiosity about the intriguing ways of the apians. Certain enigmas are acknowledged and allowed to remain opaque, never being reduced to the simple mechanics or instructions of "instinct." And at no point does Maeterlinck consider humans to be existentially superior to bees. Indeed, he utilizes a good percentage of his inkpot on reminding his readers that we are in many ways quite the opposite. (Provided we adjust our lenses somewhat, beyond our usual self-serving metrics.)

The book begins with the premise "Beyond the appreciable facts of their life we know but little of the bees. And the closer our acquaintance becomes, the nearer is our ignorance brought to us of the depths of their real existence." So to say, no matter how many details we learn about these fascinating creatures, we edge no closer to the true "spirit of the hive": a numinous and irresistible force, communicated only among its inhabitants, and perhaps not fathomable even to them. Granted, Maeterlinck is not entirely consistent in his portrait of "the waxen city." On one page he will say, "In the heart of the hive all help and love each other. They are as united as the good thoughts that dwell in the same soul." Yet in a following chapter he will claim that "each hive has its own code of morals. There are some that are very virtuous and some that are very perverse." On balance, however, it is the author's firm opinion that "It would not be easy for us to find a human republic whose scheme comprised more of the desires of our planet; or a democracy that offered an independence more perfect and rational, combined with a submission more logical and more complete."

Regarding the latter aspect—submission—Maeterlinck observes that the bee is, above all, "a creature of the crowd." Every individual instance of bee-hood submits to, and is likely physically engulfed by, the interests of the swarm. Even the regal queen, who reigns supreme in any given community, is subject to her prodigious and restless fertility and her own chronic obedience to the spirit of the hive. Bees, in other words, belong to an "almost perfect but pitiless society . . . where the individual is entirely merged in the republic, and the republic in its turn invariably sacrificed to the abstract and immortal city of the future."

Does this ring true to you, dear bees? Would you agree that you care for nothing so much as *the honeycomb to come*? Maeterlinck paints you as perhaps the most future-directed creatures on Earth, sacrificing even the queen in order to ensure the glory of tomorrow: a "future society, which the bees would appear to regard far more seriously than we."

November 16, 2019

The copy of *The Life of the Bee* that I'm reading was published in the 1950s and is currently on loan from the New York Society Library—the oldest library in the United States. Inside the front cover I was startled to find a bookplate depicting an elegant woman, standing upright in front of a large bookshelf, dressed in what I guess to be Georgian attire. This fair maiden is hand-ing a book to a Native American male, who—in sharp contrast to her honorable verticality—is cowering in a half-human, subservient way, extending his bare arms to gratefully and ceremoniously receive the tome. A slogan in Latin hovers above the bookshelf behind the two figures: emollit mores ("learning humanizes"). I admit that this image is not the most auspicious gateway to my reading experience. And it's astonishing to me that in this year of our Lord 2019 such iconography can still be circulating in the metropolis without comment or censure. Then again, it is always a useful reminder that all our endeavors in the rapidly aging New World are built upon the legacy of colonialism. Indeed, we would do well to gaze into this ugly mirror as often as possible, lest we forget this inconvenient fact. (And you can see how Maeterlinck's formal style is already infecting my own tone.)

While conducting some general research on the author—who won the Nobel Prize for Literature in 1911, a decade or so before being accused of flagrant plagiarism—I learned that Maeterlinck was asked by none other than Samuel Goldwyn to suggest a few scenarios for possible films. Apparently, Maeterlinck based one of his pitches on *The Life of the Bee*, prompting Goldwyn to toss the document aside, exclaiming, "My God! The hero is a bee!" (We can perhaps consider it a sign of progress, then, in the age of Pixar and DreamWorks, that this proposition is now possible, albeit not necessarily profitable.)

November 21, 2019

Good morning, dear bees.

I'm still working my way through Maeterlinck's book, which does indeed make your abode—despite all the toil and occasional flashes of violence—sound like an enchanting place to live. "No living creature," he writes, "not even man, has achieved, in the center of his sphere what the bee has achieved in her own; and were someone from another world to descend and ask of the earth the most perfect creation of the logic of life, we should needs have to offer the humble comb of honey." The author's account contains some lovely descriptions of the organic infrastructure of the combs, themselves created by the bodies of the bees themselves. The wax is "immaculate . . . has no weight," and is "seeming truly to be the soul of the honey, that itself is the spirit of flowers." The walls of the hive are beautifully described as "a motionless incantation." As for the honey itself, this is a "kind of liquid life" or "liquefied light," circulating through the structure "like generous blood." The whole city sounds much warmer and more intimate than our own concrete canyons.

Maeterlinck even goes so far as to compare the primary product of the hive to the main ingredient of human endeavor, for "just as it is written in the tongue, the stomach, and mouth of the bee that it must make honey, so is it written in our eyes, our ears, our nerves, our marrow, in every lobe of our head, that we must make cerebral substance." He follows this fanciful comparison with a well-placed qualification: "[N]or is there need that we should divine the purpose this substance shall serve." I find this last point quite soothing, as I find myself writing for a species that cannot read. Perhaps we humans think and reflect and speculate and write not for any specific entity or design, but simply to be able to continue doing so, just as the making of honey may not be solely to sustain a given bee population but to embody the meaningless ecstasy of its inherent sweetness.

November 22, 2019

Speaking of sweetness, what Maeterlinck calls "the gladness of June" is but a dim memory now. As a late Victorian, my Belgian guide is partial to anthropomorphism, which tends to be frowned upon these days. I sometimes wonder, however, whether we deny the animals their own passions and pleasures when we assume they are merely prisoners of "the overwhelming indifference of Nature" (to quote Werner Herzog). For Maeterlinck, however, you bees "are the soul of the summer, the clock whose dial records the moments of plenty." You are "the untiring wing on which delicate perfumes float; the guide of the quivering light-ray, the song of the slumberous, languid air." (Or as an old folk singer used to amusingly croon, about your kind: "It's milk and honey . . . without milk.")

Today, as I make my breakfast, dear bees, I will spread some blossom syrup over my toast, in honor of your now-past "festival of honey," which is also "the only Sunday known to the bees."

November 30, 2019

My apologies, dear bees, for the little break since you heard from me last. Or perhaps you did not notice my week away? The semester became an avalanche of paperwork, as it always does. The Thanksgiving break, however, has allowed me to clamber back to the surface and take a few deep breaths. Not being American, I don't really understand the combination of homeward pilgrimage, animal sacrifice, and Oedipal dread that this holiday represents. But I *am* happy to take advantage of an excuse to enjoy a feast. And the city becomes so quiet and peaceful, emptied of half its inhabitants, who now find themselves back in the shabby bedrooms, diners, and malls of the suburban or small-town childhoods that they like to think they have definitely escaped. Yet this marking of time is something I admire about my now-adopted country, and would miss if I were to leave—the clear rhythm of the calendar year, punctuated by rituals and tethered to the seasons. In my Antipodean upbringing, we had no equivalent, really. Australia Day is little more than an excuse for sizzling sausages and watching some desultory fireworks explode low over some rusting fishing boats. Christmas, for its part, is a family occasion in which any kind of beer-soaked sentimentality is soon burnt off by the blazing sun, and the industrial coconut scent of various waterproof lotions applied to restless, sand-prickled limbs.

In class last week, we discussed the "deritualization of society" and the disorientation, in both time and space, that this can produce, especially for social creatures like ourselves that experience less and less symbolic initiations or rites of passage. I asked the young folks what kinds of rituals they might be creating, unconsciously or not, to help rebond with their fellows, now that the traditional social glues have started to unstick. The three answers offered were "brunch," "SoulCycle," and "Coachella." Of course, my heart sank. But perhaps this is indeed where we are now, as a culture: inventing new rituals around consumption, lifestyle exhibitionism, narcissistic performance cults, and compulsive, joyless hedonism. (Do I sound jaded, dear bees? How can one avoid becoming so, while still paying attention to what's going on? . . . . I don't ask this as a rhetorical question. *I would really like to know*.) So to say, I fear we don't have the communal will to erect a new sabbath against the pulsing dark vortex of Black Friday.

Even so, I still hope we can do better than *this*, as the world-as-we-know-it begins to disintegrate, literally and symbolically. Some might object: Why create new rituals if we're going to be fighting over water and food soon? But that's precisely why. If any shred of the social contract is to survive the coming reckoning, we will need to see and understand one another as leaves connected to the same diseased tree, or droplets in the same polluted ocean. We need new pagan festivals, passionately designed to thank the sun for its bounty, the moon for its secrets, and the stars for their mischievous counsel. Were we to adapt a fresh ceremonial relationship to the Earth, and to one another, perhaps we would, eventually, become better shepherds of the future. Surely a collective acknowledgment of shared finitude—as with Mexico's Day of the Dead or Japan's Night of Hungry Ghosts—would encourage less selfish, heedless, fleeting fixations?

Conversely, I like to think of new rituals that celebrate life's forever bursting anew. For instance, a global rite, dedicated to marking the seven-year cycle in which all our body cells have regenerated, creating a completely different person (biologically speaking, at least). Instead of birthdays—which, no matter the cheeriness of the decorations and the colorful frosting on the cake, leave the taste of fading mortality—this occasion would celebrate the rebirth of the person. Memories that still sting will slough away, as we formally recognize the fact that such recollections, in a very real and physical sense, happened to someone else. After all, not a single cell in our flesh or bones were witnesses to an experience more than seven summers ago. What a relief! What a load off! And to mark this rebirth—this material and spiritual rejuvenation—we could build a great vessel, symbolizing the Ship of Theseus, whose timbers were repaired and replaced so many times that eventually nothing of the original was left. (Could Theseus himself confidently claim this structure to still to be his original vessel, given the constant churn of its elements?) And we could build this ship out of old photo albums, bank statements, dental appointment reminders, and love letters, addressed to a person who, at the strike of a match, no longer exists.

The seven-year itch, they used to call it, without realizing this cycle marked an almost insectoid instinct to shed one's shell-like skin.

Which makes me wonder, do you, my dear buzzing ones, have any kind of rituals or celebrations? Do you ever disentangle the waggle dance from the pragmatics of feeding and partake in a ceremony of royal solemnity? Do your sweet catacombs become a temple at each harvest moon, to observe a sacred fast or feast? A Ramadan or Passover of the bees?

June 24, 2020

Hello again, my buzzing ones.

I'm sorry that it has been so long between visits. Six months, I see! . . . A *lot* has happened. To say the least.

(And yes: I realize that you are not the same bees I was speaking to before, given that the lifespan of your kind is no longer than two months. Very possibly I'm now telling this latest news to the great-grandchildren of my former interlocutors.)

In any case, a lot has happened.

And for once, this isn't a parochial exaggeration. The whole world—the whole human world, at least—has slipped off its axis. A giant wrench has fallen from the sky and sabotaged the machinery of our lives, all over the planet.

Do the animals speak of this? Surely, they must know something has changed. Or they must at least *sense* a great difference in things. A couple of months ago, everything was suddenly so quiet. No planes. Hardly any cars. Hardly any people outside. And then the sirens. So many sirens, shrieking through the strange, late-winter silence like scissors through a bolt of dirty silk.

What has changed?

Well, a deadly virus has emerged from the shadows, and the pandemic is raging through the human population like a wave of tiny termites through a lumbering forest. The epicenter of this new plague, for several months, was none other than our shared city: New York. (What is it about this particular metropolis that makes it a magnet for apocalyptic situations, I wonder.) The first few weeks of this emergency were like nothing I've ever experienced and hope never to experience again. The dread sense of being caught inside an actual nightmare from which it is impossible to shake oneself awake; or of having stepped through the screen of a near-future dystopian horror movie, perhaps never being able to return to one's comfortable seat. (I'm aware, dear bees, how clichéd these analogies are. But at the time, in the grip of the first elongated wave of panic, this was all I could summon, since it was—it *felt*—undeniably, asphyxiatingly apt.)

Never in my most baroque fantasies did I think I would ever walk through the middle of Times Square—in the middle of the day, in the midst of a working week—and see only a handful of souls, scurrying around like frightened rats. All the stores were closed, and yet the giant quasi-living billboards continued to loop their gargantuan commercials, surreal and looming representations of a world suddenly vanished. The models posed and smirked down on empty streets, with no one to witness and covet the products on offer, save for a small group of messianic zealots who looked like extras in a forgettable Hollywood movie about The End of the World, brandishing homemade signs and shouting about the Rapture through a megaphone.

Why was I outside with the forsaken ones, you ask?

Well, I soon realized I had to recover some valuables and other items from my office before the college closed down along with the rest of the city. This entailed an epic walk, like something out of *Ulysses*, in both directions, double-masked and sweating as much from fear as from exertion. Strangers coughed in abandoned doorways. A couple of entrepreneurial figures lurking near card tables on the street were selling masks and hand sanitizer, the latter being especially hard to find after a mass rush on the allegedly salvaging gel. As I continued on my way, I was obliged to step around a slumped figure who had apparently repurposed a vintage World War I gas mask, perhaps scavenged from a nearby Army supply store.

Here in the geographic belly of the capitalist beast I indeed experienced a sublime kind of terror, a vertiginous epiphany. For it's one thing to know that our system is a thin expanse of opaque cling wrap stretched over the abyss of the Real. But it's another to find oneself falling through the plastic and glimpsing the void broiling below. (A visceral lesson one can learn only through experience, during times of war, famine, natural catastrophe, and, indeed, plague.) Suddenly, in a matter of days, the whole world seemed to have ground to a halt. And yet these shimmering shadows—these pixelated puppet shows—continued to writhe and beckon, twenty stories high.

After that surreal odyssey through an infected city, I cowered inside my apartment like a fearful limpet inside its shell. The Internet became an extended nervous system for nerve-wracking accounts: makeshift tent-hospitals being hastily assembled in Central Park . . . even graves being dug in green spaces around the city to accommodate the dead. We heard reports of a largely abandoned subway system, filled with the wretched, lost, insane, and virally vanquished, riding back and forth in carriages otherwise emptied; mobile hospital wards with no doctors or nurses, some even becoming makeshift morgues, shunting corpses back and forth between a forsaken Coney Island and Norwood, South Ferry and Van Cortlandt Park. The only time we dared make any noise ourselves—rather than tip-toeing around our claustrophobic abodes, hoping the virus wouldn't hear us—was at 7:00 P.M., when we whooped and hollered out the windows and used sauce-pans as makeshift gongs, the sounds clattering around the neighborhood. (We did this both to recognize the medical workers, bravely struggling to keep our friends and family alive, but also to simply register the fact that we ourselves were still here, though for how long no one could say.)

Of course, we humans—unlike your kind, I assume—are forever burdened with the knowledge that we are mortal and will one day expire. Our species, howev-er, is also ingenious and unique in inventing all sorts of ways and devices of distracting ourselves from this bleak and inescapable fact. Indeed, you could argue that everything we do is but one of the seemingly infinite techniques we have for banishing this dark knowledge from the front of our minds to the hidden corners in the back, where all the terrors of conscious existence lurk, mutter, and multiply.

So you see, my dear bees, I had a valid excuse for not coming to visit you for a while.

What has happened in the meantime?

Well, as the weather began to warm, and the wave of deaths crested and ebbed, the surviving New Yorkers allowed themselves to exhale. They did so, however, only in furtive and qualified ways. We are now obliged to wear masks when venturing outside, and we view our neighbors with a new and intense kind of anxiety and suspicion. "Social distancing" is the new norm, so that all interaction beyond the family is highly discour-aged. The people with some financial breathing room (who also mostly happen to have white skin) work from home, while the people living paycheck to paycheck (who mostly happen to have darker skin) work in the streets, in trucks, in supermarkets, and in warehouses—forced to contend with the capricious whim of the virus—so that the privileged people (like me) can keep eating, drinking, and watching their distracting little digital stories. In New York especially there are simply two classes now: the deliverers and the deliverees.

(I wonder in passing what your kind thinks of this strange distinction, where skin color can determine so much about the size, scope, and texture of one's life. Besides the royal hierarchy, is there a type of prejudice between different types of bee? I understand that worker bees can be prioritized over drones and that surplus males will often be physically evicted from the hive during the winter if resources are low. But are there subtle aversions—even forms of violence—based on size, color, or texture, even within these classes?)

In any case, we have a whole subspecies of human dedicated to making sure the unspoken line between the light- and dark-skinned remains unchallenged and unbroken. And we call this subspecies "the police." They have in turn invented something called "the thin blue line," which many light-skinned people believe in, and pray to, as if it were a religious conviction. The police have a terrible habit of using this blue line like a garrote to figuratively strangle dark-skinned people they take an exception to, for the audacity of simply existing, for instance, and trying to keep doing so, in public. "Good luck with that," grin the police, as they attack the breathing passages of these startled souls with a ruthless efficiency matching that of the virus.

Finally, after an especially disturbing example of this macabre reflex was circulated through the networks linking our hives, the people decided together that they had had enough of this grim sport, and—despite the ongoing threat of plague—flooded the streets with powerful and righteous indignation. Perhaps you heard the din? The police predictably responded with redoubled violence, sometimes even against the lighter-skinned people, for daring to step across the invisible line and attempting to erase it in the process. People were blinded, injured, and some even killed, in the pitched battles that spilled out into the streets.

This was a week ago, and now the city has returned to a hot and humid semblance of calm, punctuated by illegal fireworks and the random howls of the freshly traumatized. The summer solstice has passed, and yet people are unsure how to behave or how to breathe properly without outdoor concerts, bustling streets, bared bodies, and languid parks. The sky is largely free of jet-trails and Hamptons-bound helicopters, which is a blessing. America has been cut off from the rest of the world. There are no tourists. Nor can we become tourists ourselves. The death toll in New York is slowing down, but the rest of the country is learning what we went through a few months ago. There is no sign of these microscopic termites losing their hunger. No sign of the invisible fire running out of fuel. (It's disorienting to realize that from the perspective of the virus, humans are merely the landscape in which it thrives.)

During this time, we soothed ourselves with the notion that "nature is healing," with our toxic trade and polluting ways largely suspended. We heard of dolphins in Venetian canals (a hoax, it turns out), coyotes in our back gardens, penguins in malls, and other animals roaming free in depopulated cityscapes.

What do the bees think of all this? Do you feel as if my kind is finally, belatedly, being punished for our sins against you and your creaturely kin? It's hard not to come to such a conclusion. And it's hard to deny that we deserve it.

In any case, yes. A lot to convey. This is just the bare outline, really. Like everyone else, I have been on a mental and emotional journey, even as I have been tied to a single spot, like a terrified tetherball. The lockdown has done all sorts of queasy things to our sense of the space-time continuum. I may tell you about that later, when I've managed to make a bit more sense of it all. Suffice to say, I've watched the movie of my life over and over again, in the projection room of my frightened mind. (They say that just before you die, your life flashes before your eyes. But what they don't tell you is what it *feels* like when you have no idea if you're going to die or not—for months. In that case, you are trapped and strapped in front of the projector, like that character in *A Clockwork Orange*, and obliged to reckon with your own story, your own trajectory, and the way it has intersected with others . . . other people, other places, other possibilities.)

So again: While in some ways I have nothing to report—in the sense that nobody is really *doing* anything anymore—I am also obliged (like many others) to acknowledge that perhaps I haven't really been doing anything for a long time now. Perhaps I've been doing an especially exhausting and frenetic type of nothing my whole life. Perhaps almost all of us have.

A realization that is a lot to process.

But now that I've finally come to see you again, to tell you what's been going on, I hope I'll get back into the habit. (And I apologize in advance if this is too much, or too boring, to listen to right now. I'm sure you've got your own problems, your own issues, your own crises to deal with. In which case, feel free to tune out. Hopefully the sound of my voice, cleaned of all content, can at least be somewhat soothing to you, as you go about your daily business, as your collective honeyed hum can be for me.)

June 25, 2020

Hello, buzzing ones. I'm back again.

When I promised I would get back into the habit of telling you things, you didn't quite believe me, did you? And yet here I am.

Today happens to be my birthday. I am now entering the last year of my forties. I originally planned to be celebrating on the French–Spanish border, near the Atlantic coast, but given the circumstances, I am spending today the same way I have spent every other day for much of the year thus far—sitting in front of my computer, clicking back and forth between virtual windows.

But I'm still here. Still breathing. And for that, I'm very grateful.

Hopefully I'll see the Old World again soon, even as the European Union prepares to announce a ban on all flights from North America. It's strange, though, not to socialize with anyone outside the home, except through a screen. Talking to you, my insect friends, is the closest I've come to unmediated intersubjective contact for months and months. (And you'll notice I'm keeping six feet away, just in case.) It's frustrating not being able to organize a picnic, or jump onto a train. (Few things make my spirits yearn as the sound of a train whistle.)

One particular thing the virus has robbed us of is the palpable promise of an Elsewhere. And I never realized how important such a promise is to one's sense of place. It is as if the *here* needs a strong understanding of *there* to really realize itself. Moreover, we're also losing the sense of an Elsewhen, as time dissolves into an eternally circulating, yet increasingly stagnant, limbo. Each day melts into the next, in a domestic version of Nietzsche's eternal return. Each moment could be swapped out with any other, and the day would not feel stuttered. Likewise, it is as if the *now* requires both *the then* and *the to come*, in order to manifest its nowness. Its now-itude. (Perhaps you can see how such disorienting spatial and temporal conditions secrete their own warped, insular, and sterile speculations as the space-time continuum continues continuouslessly.)

Each dawn is thus a promise that can never be kept.

So until I can actively launch myself toward such Elsewheres, I will concentrate on cooling my heels, remembering Elsewheres past and daydreaming about Elsewheres to come.

July 14, 2020

A New York memory, from six years or so ago: I was dog-sitting for a colleague, who had been kind enough to lend me her chic apartment in a Clinton Hill, Brooklyn, brownstone while she was away for the summer, when I found myself "between accommodations." This neighborhood had once been middle-class but then changed color and flavor as a result of "white flight" in the 1970s and '80s. Now, the white ones who had fled were flooding back into this part of Brooklyn, bringing their yappy dogs and yoga mats with them. Six years ago, the transition was not complete, so it was an interesting exercise in embodied gentrification to walk a floofy white poodle through the streets, still populated by many Black folks.

One morning I tried to speak with the tenant in the apartment on the top floor, something to do with renovations and requested access to the ground-floor back garden. As I stood on the landing of the third floor, buzzing the doorbell, a sudden voice from right above my head caused me to nearly leap out of my skin. I looked up, and a man's head was sticking out of a recess in the ceiling. This inverted jack-in-the-box looked at me with a suspicious expression.

"What do you want?" he asked, defensively. The man was around forty years old. He did not look homeless, but neither did he seem particularly homed, peeping out of a ceiling like that above the brownstone's internal stairwell. I tried not to connect this unexpected moment with half-remembered scenes from *The X-Files*.

"I'm trying to contact the people in 3A," I answered, as casually as I could, my head bent back nearly ninety degrees.

"And why is that?" the man asked, in his wary way.

For a moment I wondered if I should even be having this conversation at all. But instead I answered, "They left a note, wondering if I could give their electrician access to the meter in the back garden."

The man looked me down and up, and nodded, as if to acknowledge that he believed me.

"They aren't home at the moment."

"I see," I replied.

"They should be back a bit later," he added.

"OK," I said, already retreating down the stairs. "I'll try again later."

I then returned to the ground-floor apartment and made sure to lock the door. Later that evening, the tenants of 3A—a Midwestern-looking couple in their late twenties—knocked on my door and repeated their request, this time in person. Access to the garden was arranged for the following day. Before they returned upstairs, I couldn't help but ask:

"So . . . who is the guy . . . ummmm, living above you?"

The couple exchanged a pinched look.

"We're not entirely sure," answered the husband. "We've only been living here a couple of weeks now."

"But it's not just the man," added the wife, in a stage whisper. "There's a woman as well."

"It's like having a ghost in the building," said the husband. "He knows when we're in, and when we leave. We can hear him walking above us and watching TV."

"I figure he thinks the attic is his apartment," added the wife.

"We complained to the landlord," said the husband. "And he promises to get him out soon."

"Well, he certainly gave me a shock," I admitted.

"It's all a bit much to think about, given the baby and all," said the wife, patting her tummy bump.

"Of course," I said, not sure exactly what she meant, beyond a middle-class sense of discomfort with illegal squatters living in the walls.

A week or so later, I watched guiltily from behind the lace curtains as the man and his partner were evicted. The now-ejected squatters sat on a stoop across the street, smoking and spitting as they watched a team of men load their humble furniture into a truck. The little white poodle that had become my charge yapped at all the activity outside as the couple sat on the stoop with watery eyes, just staring back at the building. The man clutched an Army bag, and the woman had a plastic tricycle slung over her shoulder. Even after they eventually summoned the will to stand up again and shuffle down the street—toward who knows where—the poodle kept yapping and yapping and yapping.

July 15, 2020

Recalling this mystery man living in our ceiling dislodged another New York memory, this time from the very beginning of our move to the Big City. After a few months in a sublet on Clinton Street, on the Lower East Side, my wife and I needed a new place from which to launch our campaign for fame and fortune. I can't recall what dubious networks we used to find an affordable place, but it was more likely craigslist than a legitimate real estate agent, since the apartments we looked at are still burnt into my memory, given their almost noble dishevelment. Indeed, "bottom of the barrel" only begins to describe the inventory on offer, including one studio apartment with the small double bed literally pressed up against the fridge, and another with a full bathtub in the kitchen area. The most memorable of all, however, was the place shown to us by a classic former street urchin who limped rather performatively around and mumble-trilled in his own personal language, like a human–pigeon hybrid. The apartment was grimy and dark and looked like a thousand other hovels in the neighborhood, perversely boasting "original condition." This place, however, had a unique feature—a large hole in the ceiling. This startling puncture must have been relatively recent, since there was still a large pile of debris directly underneath it. The ceiling of the apartment above could be seen through the cavity, though there was thankfully no sign of anyone living up there during our brief visit. The whole scene looked like the aftermath of some kind of *Roger Rabbit* human–cartoon mishap. Our mumbling-trilling guide did not even blink as he effortlessly failed to acknowledge the large flaw in his merchandise. As we were shown the abundant electric outlet options, and obliged to watch an elaborate demonstration that the toilet did indeed flush, we nodded politely, somehow feeling obliged to play along. Somehow it seemed important that we also avoid mentioning the elephant-sized hole in the room.

Needless to say, we hustled off to our next (fictitious) appointment as soon as possible. And I still wonder if we were the victims of a hidden-camera-type prank. Then again, the whole thing was so deadpan, and so in keeping with the neighborhood, that I fear some poor runaways, somewhat more desperate than we, ended up living there soon after, with a modest reduction in rent for the "inconvenience" above their heads. (Which could, after all—as their slumlord probably insisted—be considered a bonus skylight.)

July 16, 2020

A sudden flood of memories from the same period.

We were subletting a tiny one-room apartment in a quintessential Lower East Side building, pre-gentrification. (The kind in which cockroaches would scatter in a perfect circle, when the unwise decision was made to open the oven door.) Our bed was two floors above a social club for aging Dominicans, who would flirt and fight and dance on the street to loud salsa until 2:00 in the morning. Our immediate neighbor snored so loudly—and the walls were so thin—that his sinus issues would keep us awake. When God was feeling merciful, however, the neighbor would spend the night in a fetal position, out in the corridor, in front of the door of his apartment, until his wife decided that the alcohol had worn off. On the other side of the opposite wall, middle-aged Chinese women would play *mahjongg* and laugh until late into the night. This wall was also more of a spatial suggestion than any kind of sonic barrier to muffle the clacking of ancient tiles. (At least one of these women would also use the floor just outside our door as an extension of their chopping board, when they had bought too much poultry to fit into their kitchen.)

Two Chinese latchkey kids—young girls around seven and nine—would rattle around the building after school, looking for something to keep them occupied until their parents came home from work. I recall one occasion when these urchins followed us up from the street so they could poke their little curious heads through the front door of our tiny apartment.

Neither of them seemed to draw breath as they took turns peppering us with questions and unsolicited observations.

"Are you married?" asked the younger one.

"How much money do you make?" asked the slightly older one.

"How much rent do you pay?"

"What did you just buy from the grocery store?"

"Is that a futon?"

"Is that also your bed?"

"Where are you from?"

"I like that picture."

"What's your favorite food?"

"Our family hides money in the watermelon."

After answering their queries as best we could, we made them promise never to repeat that last piece of information to anyone else ever again.

July 18, 2020

There is a new comet in our cosmic vicinity. It has the romantic name C/2020 F3 (NEOWISE). They say it is one of the brightest in many years, visible by the naked eye, toward the North. But I have yet to see it. (I fear it may be smudged above the horizon, behind the large storage building between my apartment and the Northwest quadrant of the sky.) Historically speaking, comets have often been considered omens of disaster. (*Disaster*, of course, means, "bad star.") Surprisingly, few people today seem to be aware of the remarkable coincidence of such a spectacular comet's visiting us during one of the most globally catastrophic moments in a century. Is this because paranoid conspiracy theories have supplanted good old-fashioned superstition?

July 19, 2020

They say that beauty is "the promise of happiness." But I'm not so sure. If classical Hollywood taught us anything, it's that beauty is the promise of ruin. Of humiliation. Or, at best, of gentle rejection. In my view, *happiness* is the promise of happiness. For surely one is most happy when there is the strong chance of more happiness around the corner. (A feeling that may or may not be tangled up with the idea of beauty.)

July 20, 2020

Even when we ourselves are middle-aged, we wince at the visible signs of the accelerated aging of our parents. We find ourselves avoiding looking too closely at the wrinkled hands, the crinkled faces, the clouding eyes of our beloved progenitors. Instead, we project a reasonable facsimile of an idealized parental image—neither old nor young—onto the naked visage that faces us.

(An existential jolt that one would not feel so violently if sharing the same place as one's parent or parents on a daily basis.) And yet, it only stands to reason that the parent, too, is shocked by how old we, the former child, has now become. We look weary. We look besieged, anxious, and worn. We have hardly any hair left. Or too much, frizzing out in unruly directions. Where once we were a glowing and creaseless creature—a wriggling babe, plucked fresh from the larvae field—we are now to be counted among the bent and nearly spent.

July 21, 2020

The pandemic rages out of control. It's maddening. The flashpoint, at least here in the United States, is the wearing of masks. While the Centers for Disease Control and Prevention insists that we could get the virus under control in six weeks, if everyone wore a mask outside, millions of Americans feel that this is too much of a sacrifice to make, when it comes to the incoherent gods of "freedom." Yesterday, for instance, I felt the urge for a coffee, as I hadn't had one in months. As the young (masked) barista made my espresso, she told me about a customer from earlier in the week who had approached the counter, mask around her chin. (A pointless and potentially lethal nonchoice, made by every third person on the street at present.) When the barista asked this woman to please put her mask on properly before ordering, the customer said, "Why? There's no one in here."

It's rare that such a short statement can say so much about the state of the world, and about the unthinking, unblinking way in which so many people move through it. Meanwhile, the infection numbers in Australia continue to be tiny, in comparison to those in North America. But a lockdown has been reinstated Down Under because the security guards charged with overseeing the quarantine there kept having sex with recent returnees from overseas. (Some of whom were indeed carrying the virus.) This is Australia in a nutshell. Quasi-fascist, but also compulsive and hopeless.

Elsewhere, the UK swiftly revoked a law obliging the wearing of masks when ordering food-to-go from a restaurant after one of their high-ranking ministers was caught on camera doing precisely this, unmasked. In other words, the government of Britain would rather protect one privileged idiot than the population of the rest of the country.

The United States, the UK, and Australia.

Or as I now call them, The Three Stooges.

July 23, 2020

Trapped at home, I try to live vicariously through the birds that crisscross the sky in Manhattan. Now that hardly any planes fly over—and only the occasional police helicopter blights the sky with our human clamor—the birds have the above-realm pretty much to themselves. (And now that my neighbor kindly lets me use his deck while he's in lockdown at his much larger property on Long Island, I can lie on one of his canvas sunbeds and watch the avian traffic moving back and forth across Manhattan Island, between me, marooned on this craggy, air-conditioned sandstone cliff and the hazy clouds in the lower atmosphere.)

The cormorants are the most focused and determined of the local feathered commuters. They fly from West to East, or vice versa, in groups of two or three, with no sense of anything but impatience to get to their destinations. Perhaps there is time to dawdle or explore later, but not as they fly across Manhattan. The seagulls, in contrast, are not so driven, and they often wheel around in small groups, way up high, following some complicated and looping expressway that only they can see. You will never spot one of their squawking kind anywhere other than by the rivers on each side of the island, which is notable, given how narrow this compressed atoll actually is. They clearly abhor being more than a few wing-flaps from salt water. The geese like to make their famous racket, as if complaining about the route even as they fly it. (Usually North to South, or vice versa, in contrast to the cormorants.) One time I even saw a flock of geese realize mid-flight that they were going the wrong way—or else they simply decided to turn back for some reason—and lean into a collective U-turn. The ducks often mimic the retro-kitsch wall decoration found in so many homes in the 1960s, by flying in threes. The mourning doves hop from building to building, their original grief no longer weighing so heavily upon them. Each time they launch themselves into the air, their wings make an unnatural sound that brings a smile, as if they are in fact wind-up toys or clever automata. The starlings flutter in small gangs, alighting on terraces and water towers, ready to even steal the food out of someone's mouth, if necessary. (Sadly, I've yet to see them Rapture themselves into murmuration over the rooftops here.) Ravens fly together in an "unkindness," often harassing the raptors with their unsettling caws. Kestrels usually fly solo, like diminutive but deadly arrows, returning with dead rodents for their mates, among the rusting fire escapes. The red-tailed hawks and peregrines tend to circle high above the buildings, sometimes casting shadows as large and as looming as an apartment block. Occasionally, I can see a loping egret, flapping its wings in a rather languid manner, hovering between dignified and awkward, seemingly defying the laws of physics. I half-expect them to consult a fob watch, retrieved from deep in their feathery chests mid-flight. The even more rare crane looks like an escaped hieroglyph in search of the ancient Egyptian obelisk on the East side of Central Park. Most common of all, however, are the pigeons. Familiarity usually breeds contempt. And yet I see a beauty approaching grace in some of their swooping vortices, their wings shaped like an iridescent boomerang. When the seasons are changing, they gather in large flocks and practice complex choreographic maneuvers together above and between the artificial canyons of the Upper West Side, as if preparing for an approaching aerial war. Given all this discipline, I suspect that the pigeons will in fact win such a war and inherit the Earth, or what's left of it.

July 31, 2020

Slightly cooler today, but low clouds and oppressive humidity. Not exactly a reprieve from the heat wave that has been sloshing back and forth across the country these past couple of weeks. My fancy new fridge could not handle the intense temperatures and gasped into a dysfunctional state for a few days. Hopefully, though, after some technical tinkering, it's starting to behave itself again. Our super insisted that I install the air-conditioning unit more securely, to avoid its possibly falling and squashing his little yappy dog, so that was an additional $270 that I simply kissed goodbye, hardly getting to know it in the first place. (Something I can ill afford, now that all our salaries have been cut in a bid to keep the college afloat.) Worst of all, the new neighbors (a curse upon them!) have taken this opportunity to do a "gut renovation," which means that drills and jackhammers, and a variety of other power tools, are making my whole world vibrate. I can feel it coming through the wall, just inches from my ear, and up through my feet and fingers. My computer screen wobbles in resonant sympathy with the hideous racket, like a bully that sits on your head and farts in your face while saying, "Stop farting. Why are you farting? You're farting on your own head, loser!" I fantasize about giving my neighbors a gut renovation as soon as I see them.

(This makes me wonder, dear bees, are you ever bothered by construction noise in the combs? Or do your hexagonal quarters essentially assemble themselves, cushioned on absorbent pillows of wax?)

Normally, I would escape to my office, or to a café or library. But all these things are still closed. (Outside, I would say, 80 percent of the people on the street are now wearing masks, although many still seem to find the concept challenging, especially when it comes to covering the nose.) I shouldn't complain, however, given that nearly 50 percent of New Yorkers are facing imminent eviction now that the economy has completely tanked and neither local nor federal politicians are doing much to help. (In contrast, as usual, to the rest of the world, which is subsidizing lost income and covering rent shortfalls.) America braces for another tsunami to hit, this time a massive wave of people violently and suddenly unhomed, unwaged, and uncovered by health insurance, even as the pandemic rages on. The whole thing is too bleak to fathom. But like the footage of the actual Japanese tsunami in 2004, this tide doesn't appear so threatening at first. Yet it builds and builds with such relentless force that the surge sweeps away everything in its path: cars, trees, houses, and, of course, people. Perhaps this is part of the problem. Hollywood has conditioned us to acknowledge disasters only if they are spectacular, visible, sublime. This current catastrophe—like climate change and environmental devastation itself—is essentially unrepresentable, what some have called "slow violence." And as a result, the human mind cannot fully believe in it. The apocalypse must be televisual, like the September 11th attacks, or else it is too easy to disavow, deny, or distract ourselves from.

When I look out my bedroom window, however, the reality hits home, in a very literal sense. I've already seen one family vacate. And now, in the apartment below, I can see the woman who lives there, packing up all her belongings into boxes. It all looks so forlorn. Especially at night, as the light shines mercilessly on the banality of tragedy. Edward Hopper himself would weep.

What infuriates, of course, is how easy this would have been to avoid. But this country is, for whatever deep symptomatic reason, in some kind of gleeful death spiral, like at the end of *Dr. Strangelove*. Rednecks straddling bombs and smoking cigars are riding the apocalypse like rodeo clowns. Only they have their own soft landings to look forward to. Unlike the other 99 percent.

The future remains radically uncertain. One thing is for sure, however. And that is that the next few years are going to be rough for practically everyone. It's going to be the Spanish flu, World War II, and the Great Depression all rolled into one. (With heavy dollops of *Black Mirror*, just to make everything that much more sinister and dystopian.) So it's going to take all our mental, physical, and spiritual resources not just to stay alive, but also to stay sane. But if we simply concentrate on surviving, then we have practically nothing to live *for*. That's why people are still attempting to reclaim the streets, even at risk of their own health and welfare, to fight for something other than sheer transaction and extraction, something beyond mere bleating obedience. We may be technically alive, but we're not living in any organically rewarding way. Joy, at this point, feels like the mythical protagonist of an old ballad, one that the old Boomers sing around the backyard firepit, when their pills kick in, and the twinkle in their eye returns for a moment.

This is a question that keeps nagging at me. *What have we actually lost?* What have we lost since the virus started plowing through the population and seemingly upending every aspect of our way of life? Have we really been banished from some kind of relative Eden, where we enjoyed the innocent pleasures of true human interaction? Or has this claustrophobic situation merely thrown the hellish extent of our former alienation into a clearer light? Are we merely mourning the *simulation* of society, of a connection to "the People"? Is this "the trouble" that the philosophers insist we must go *through*, to get to the other side? Or is it merely the accelerated intensification of a dehumanizing—deanimalizing!—process that started many decades, or even centuries, ago?

August 1, 2020

August? Already? Really?

It is already a cliché—the way time both flies and crawls, during lockdown. It seems like a million years ago that we were all walking around, without a thought of face masks, in February. And yet it also feels like the year is scrolling past at a dizzying speed without any traction at all to help slow it down into a regular seasonal tempo or rhythm.

But, on a more whimsical note, I like to think of August as arriving in staggered stages, depending on where you happen to be in the world (and also depending on the caprice of the supply chain and delivery system). I'm not sure where each month is fabricated. Probably in a vast factory in Mexico or China. But wherever the point of origin, it is shipped to one's door—eventually, at least—for the most part. This year, it seems I received my August in one piece, and on time. The day is hot and humid, just as one would expect. I paid my bills online today, as I do on the first day of each month. My neighbors, however, may be still living in July, if the post office failed to deliver their monthly subscription, to each month, on time. If that particular package has gone missing, my neighbors may be obliged to live in July for another four weeks. And in that case, they will just have to hope that September arrives as scheduled, on the first day of that month, as the leaves begin to shrivel and drop.

Meanwhile, this being August already in my world, I should really start thinking about the new semester, which is looming; it's only a matter of three weeks away now. But the idea of teaching online again fills me with a special kind of dread. The university leaders insist that the show must go on, even when the audience has vanished. But in this case, the audience—our students—reappears, in tiny boxes on our computer screens. Mushrooming *Brady Bunch* boxes of boredom and despair.

The perfect conditions, of course, for a shared educational adventure!

August 3, 2020

With the premium on private outdoor spaces during the COVID crisis, I think about how little I've had access to such as an adult. I've lived in more than two dozen places since leaving home at age seventeen, and only one or two of these rented dwellings had anything resembling a garden or a terrace. In terms of internal space, or sheer volume, the most impressive of these was the apartment in Amsterdam: an old canal house, on the first floor, that had been renovated to remove all internal walls. This meant it was one vast room, stretching between the canal on Da Costakade and an inaccessible courtyard at the back—more than a hundred feet—with large windows on each end. Some heavy hidden doors could efficiently divide the room in two, and at night we would make sure these sliding doors were closed, to ensure that the rodents which gnawed our wooden spoons into strange shapes would not climb over us in our sleep. (In New York, a couple of years later, I awoke to the singular sensation of a mouse nibbling on the palm of my hand. When I realized what was happening, my entire arm jerked up violently; and I can still see the poor fuzzy creature bouncing off the ceiling, and back onto the bed, and then rebounding o to the floor, and finally scurrying away.) In Amsterdam, however, the mice remained unseen, although they could be heard long after midnight, head-butting the doors with a Flemish determination, trying to get through, and expecting who knows what on the other side.

While this apartment didn't have an outdoor area, it did provide an ever-changing view of the canal, which spooled like a piece of timeless European film across the red-brick sprockets of my apartment. On long summer evenings, small boats would putter by, carrying smiling people sipping wine around tiny tables covered in candles. The golden light reflecting off the water would ripple on the ceiling inside the apartment, creating a personal, and unseasonal, *aurora borealis*. In winter, flocks of geese would replace the boats, until the water threatened to freeze up completely. No matter the season, however, the herring man could be found on the corner, outside the living room window. The herring man was a rough and surly fellow who did not smile once in the eighteen months I watched him. He sold those briny little snacks that the Dutch liked to eat on the go—a pickled herring simply dropped down the throat, like a slimy spell against bad luck. One day I saw him trip over his own feet while moving a large tub of fish dip from the back of his car to the tiny wooden hut from which he sold his pungent wares. The herring man looked around with narrowing eyes, failed to notice me behind the windowpane, scooped up as much of the dip as he could salvage back into the tub, picked away a few pebbles and bits of bitumen, and then put it on display as always, for some unsuspecting people's lunch. The herring man had a chubby little boy who would hang around the little hut after school, bored out of his mind, throwing random found objects into the canal, and thinking nothing of dropping his shorts and emptying his bladder into the water.

From this window I watched women sometimes pedaling three children at once, wedged like piglets or ducklings inside a sturdy bucket, affixed to the front of the handlebars, rain, hail, or shine. (Mostly rain, since the North Sea still refuses to recognize Holland as a legitimate landmass, because humans built this country themselves, against the organic law of things.) I also recall watching TV when a tram went half off the rails, just outside the window. The commuters were trapped, since any move to escape the vehicle sent a terrifying electric shock through the whole tram. They were ordered to sit tight, as the authorities surrounded the vehicle, and scratched their heads. The scene felt allegorical somehow—all those weary people, trying to continue reading the newspaper, or chatting with their neighbor, knowing that they could all be fried at any moment, like one of the herring sitting in the herring man's hut, just out of reach—close enough to see, but too far away to taste, perhaps ever again. Sitting there, my fellow men and women, wide-eyed, on the other side of mortal peril.

Thankfully, the electricity to that line was eventually turned off, and the passengers stepped off, exhausted and relieved. Some even had a celebratory snack at the herring man's stand, since he had stayed open late in the hope of just such a boon. Had things gone differently, however, and the commuters had all died after being fried in a sudden moment of shared, agonizing ecstasy, I imagine, the herring man would have simply shrugged and grunted, and shut up his hut.

August 4, 2020

A radical compression. An almost violent narrowing. This can be a very clarifying thing. Like the pinhole aperture of a camera, reducing the world to a tiny point is so often the technical premise to capturing the panorama all around us. While we cannot, in our metaphysical parochialism, fathom the unfurling world in any expansive or comprehensive sense, we can, however, trick it into a tiny box and thus have some kind of control over it, some domesticated perspective upon the overall order of things. Today, during extended lockdown, the computer is our camera; our obscured room. Our computer screen is the pinhole through which we access all of Creation. But the radical compression—the violent narrowing—remains. We are trapped inside the *camera obscura* and no longer have access to the real, tactile, sensual world. Better to trade the entire digital universe, with all its infinite rabbit holes, for one small patch of actual garden, where the comings and goings of a single ladybird are of more consequence than a tweeting president or preening celebrity.

August 6, 2020

A mini-fable for you today.

A drowsy butterfly is resting on the arm of an Adirondack chair, which itself sits on an old splintery porch, not far from the actual Adirondacks. Sitting in the chair is a retired scientist, sipping on a gin and tonic. He is charmed by the proximity and beauty of the butterfly, and tells his young grandnephew, sitting nearby, of the famous theory. "Did you know," he says, "when this butterfly flaps its wings, it could conceivably cause a tiny vibration that grows and grows and sets off such a cascading chain of meteorological effects that this tiny, pretty critter ultimately causes a devastating storm in the Amazon?"

The butterfly is horrified upon hearing the scientist's words. Until this moment, it had been entirely ignorant of the awesome and fearful power contained in its multicolored wings. And so it continued to perch on the arm of the chair, paralyzed by the responsibility it now embodied. The scientist remarked to his grandnephew the following day that the butterfly had not moved.

"Maybe he's dead," shrugged the boy. But on closer inspection, it was determined that the creature was still alive, but just unwilling to depart.

"Maybe he doesn't want to create a typhoon," added the grandnephew, chomping on a peanut butter sandwich.

The retired scientist laughed.

But the butterfly remained, effectively glued to the spot, day in and day out, until the temperatures started to drop, and the butterfly itself finally dropped onto the porch, moments from perishing, secure in the knowledge, at least, that no other creature had been harmed by its thoughtless actions, of just moving through the world.

August 9, 2020

Today I bring the sad news of the death of Bernard Stiegler—a prolific and influential French philosopher and certainly a huge influence on my own work. Indeed, at least three of my books were in direct dialogue with his provocative ideas. (Especially his idea that modern technology is destroying not only our inherited knowledge and wisdom about how to live—that is, how to practice the ancient art of being a human being—but even corroding our *capacity* to love and care.) The fact that Stiegler took his own life makes his legacy even more urgent, even more poignant. He was the most courageous and detailed diagnostician of the malaise we all feel but that he called "ill-being." This did not, however, immunize him from the very same.

Meanwhile, the people of Beirut are protesting in the streets against a government so incompetent and corrupt that they left enough ammonium nitrate to level much of the city sitting idly in a warehouse for nearly a decade. This negligence was tempting fate. And fate eventually arrived, in the form of a blast that shattered millions of windows and thousands of lives.

In the United States, the virus continues to burn through the population, since so many people have decided that wearing a mask is somehow unpatriotic or un-American. (What do we call the opposite of mass hysteria? How to account for blithe nonchalance, even as people drop like flies all around?) It's reported that one out of every three people in this country would refuse a free vaccine, if available. Indeed, it's less an irony than a political kind of poetic justice that American exceptionalism is what is now leading to the ultimate demise of the nation.

August 13, 2020

I realize that I still haven't confronted the question I posed myself several days ago. What have we lost, exactly—we humans, that is—since the pandemic began? Certainly, we feel—as a society, even as a species—that we've lost something at once essential and intangible. But given how dire things already felt in 2019, what was it, after all, that remained? What precious elements of our kind persisted, despite all the compounding horror, greed, selfishness, and stupidity? What forms of being—gestures, practices, rituals, customs, arts, and so on—survived our attempts to extinguish anything and everything that did not obediently and efficiently lubricate the blind machine that vacuums up every last dime up to the top of the extraction machine?

I'm sure most of us posed Peggy Lee's famous question to ourselves—"Is that all there is?"—with more and more frequency, even long before COVID. What would it take to *truly* enjoy ourselves in a society that has traded in the deep currents of pleasure—as well as the sharp peaks of bliss—for mere *fun*. ("As seen on TV.") As Stiegler and others have insisted, in our scramble for material gain and narcissistic recognition we have lost all touch with what the Greek philosophers called "eudaemonia": a profound kind of happiness, rooted in companionship, virtue, well-being, and a general shared sense of human flourishing. (*Hedonism*, after all, meant something very different in the time of Epicurus, quite the opposite of how we think of it today, for it described the pleasure obtained from a modest and simple life in which the appetites are curbed and controlled through a conscious practice of general social satiability.)

Were I a rich man, I could travel. I could drink the finest wines and eat the finest foods, and perhaps even surround myself with the most stimulating and glamorous company that money can buy. (Or at least rent.) But if the people who facilitated this travel, who made and served me the meals, and who laughed at my jokes, were themselves alienated from themselves, then they could be little more than puppets or extras in my own ultimately empty scene. In this case, we would all just be going through the motions. There would be no *there* there, as we all suspect and feel, most of the time. (Indeed, why do we do most of the things we do? Well, "to make ourselves a little bit of money" is the answer, and also to make somebody else a *lot* of money. Otherwise, it is to distract ourselves from this stark fact. No wonder we can feel so lonely, even as we're ostensibly "just having a good time.")

And yet nowadays, in 2020, even *this* hollow automatism—the kind that increasingly chafed our souls, as beloved places and ways of relating to one another were squashed under more profitable models, more efficient business plans—seems like a kind of enchanted Eden from which we've been summarily banished. Today, the most banal sitcom or drama made before the coronavirus, appears to be a dizzying *mise en abyme* of options, vistas, and possibilities. They contain multitudes and vectors of movement and connection that seem positively heretical today.

So, I ask again: What is it, precisely, that we now—truly, decisively, definitively—feel evaporating there outside, on the other side of those forehead-smudged windows behind which we still cower, awaiting the Rapture and deliverance of a working vaccine?

Could the answer simply be "society"? After all, one of the diabolical architects of this grim future is

Margaret Thatcher, who notoriously insisted, "There is no such thing as society." Even as this Iron Lady was wrong when she made such a pronouncement, she subsequently did an impressive job at turning her words into reality. Then again, I suppose it is far too simplistic to turn our collective loss into a single "thing" or object. After all, we've lost so many things. We've lost hope. We've lost direction. We've lost collective orientation. And as Stiegler insisted, we've lost the art of knowing how to live. (Where "making a living" refers to all the things we do to make life worthwhile—eating together, working together, playing together—rather than simply scrape by.) Meanwhile, Mark Fisher's "slow cancellation of the future" has suddenly picked up speed. The optimists believe this to be a painful transition to a more compassionate and sustainable future, a shedding of the old, suffocating skin. Whereas the pessimists fear this is the last gasp of any organic resistance before the Matrix boots up into a dark kind of omnipotence in which nothing is tolerated, save that which has been designed and anticipated by the Algorithm.

Some will no doubt accuse me of wallowing in a flabby kind of romantic nostalgia. After all, as long as there have been humans there has been strife, estrangement, bitterness, corruption, and collapse. But we have frightening new tools at our disposal. Tools that we can barely contain. New weapons of control and coercion. This is patently not a case of our species' simply conducting business as usual. Indeed, if I believe in anything, it's the central tenet of critical theory: "Things weren't better before. But they sure are getting worse."

Yes, we've lost so many things. What's more, we don't even realize it, because how can you miss something you yourself never had, or experienced? We've lost ways of interacting. We've lost certain rhythms and flavors of waiting. We've lost many languages, of course, and worldviews. We've lost words, phrases, expressions, tunes, dances, scents, sounds, stories, theories, ideas, insights, institutions, habits, philosophies, skills, and forms of knowledge. We've lost fads, fashions, foibles, and follies. We've lost microcosms, milieu, specific modes of *je ne sais quoi*, as well as infinite honeycombs of ambiance. We've lost things for which I could not even hazard a name—nor about which entertain an inkling—since that entire category of existence has slipped under the surface of the known and disappeared without a trace. Historians tend to fetishize and endlessly mourn the burning of the Library of Alexandria. But countless indigenous cultures have since understood the unfathomable horror of witnessing the erasure of entire peoples, and everything they contain. (A North African proverb apparently states: "When an old man dies, a library burns to the ground," perhaps remembering the flames of Alexandria.) The white races in particular have traditionally wiped out unique ways of being, as if they were merely wiping away a smudge or a tear, even as they expunge their own essences and legacies through less spectacular forms of neglect and amnesia.

But if history is essentially a chronicle of things passing away into the past, then why pause and lament? What is so different about this particular present that we find ourselves in? Stiegler again provides a convincing answer. We've lost the capacity to *transmit* forms of being, knowing, and doing from generation to generation. Our relationship to time, and thus to one another, is so impoverished and short-circuited by impatient timescales and pressured exhortations that we can no longer create the "long loops" that allow for stories, feelings, sensibilities, and mutual understandings to be passed down, through the decades and centuries. In thinking only of ourselves, and our immediate desires and fears, we have thrown away the baton that we pass down from parent to child to grandchild and so on, whether this be a recipe for ramen, a specific way to hold a cricket ball, or a commitment to Confucian principles. As such, we've lost environments, places, moods, and spaces created by human needs, whims, desires, and dreams, as opposed to those benighted spaces—standardized, loveless, neon-lit—sculpted by nothing other than making money with the left hand and saving money with the right. (And yes, I understand the perverse kind of alienated meta-romance that such non-places afford. I am a fan of Ballard and Wong Kar-wai. But the older one gets, the more these read as compensatory spasms, amid a wider tragedy or travesty.)

How can I be so sure that we are in the midst of such global privation? Because I have been the beneficiary of glimpses of the beauty—both opulent and humble—that existed before me and that are now passed on to me in ruined or anecdotal form: lost cafés, lost gatherings, lost symbolic evolutions, lost landscapes, lost textures, lost lighting, lost odors, lost contraptions, lost emotions, lost joys, lost toys, and lost longings.

We are afforded glimpses of such, in old photographs, paintings, books, museums, and music. Very occasionally I will be touched by the quicksilver of this unfamiliar spirit, as if visited by a time traveler. Then I find myself in a similar position to Proust's famous narrator, suddenly seized with the overwhelming sensation of involuntary memory. In this case, however, the madeleine-flavored trigger does not release a flood of long-buried personal recollections but those of my ancestors, or even those of unknown progenitors, of a wholly different bloodline. How to keep such rare traces alive? Can we observe a kind of "fidelity" to everything that came before, and that then happens to cross our own path? Or is that a form of madness? Is it yet more human hubris—trying to convince Saturn to regurgitate his own children. (Hopefully before digestion!)

Sadly, most attempts to resuscitate the patient (that is to say, the past) are destined to fail. Rather they result in kitsch, or pathos—or kitsch-pathos—like those people who insist on wearing clothes from the 1920s every day, even as they catch a Tesla Uber home from the downtown speakeasy. This is the rub, I suppose: Tradition must have at least a minimum of continuity, a modicum of thread or flow, or else it becomes a literal dead end, like those forsaken, devitalized objects that fill our museums, no longer animated by the energy of use, reduced to a lacquered shell of itself—like an iridescent beetle pinned to a velvet board.

Then again, I don't want to sound too conservative or nostalgic. Things must indeed keep moving and evolve. I suppose it's a question of babies versus bath-water. Trees *qua* forests.

And when I become too maudlin about our global disorientation, regarding this shared human adventure, and our lack of collective direction, I remember the cheeky definition I read somewhere, many years ago, scribbled on a toilet wall: "Tradition is just peer pressure from dead people."

August 14, 2020

The number of people wearing masks in my neighbor-hood is now around 90 percent, even as many of these are pulled down below the chin or the nose. I have always found masks creepy, especially when glimpsed in public, as one sees often in East Asia; and I can't imagine I'm alone in that. After all, in this context they signify ill health, anonymity, infection, and the uncomfortable proximity of death. And this unease is only enhanced when almost everyone is wearing a face-covering. I understand, of course, and appreciate the rational reasons for wearing a mask. But while I'm relieved to see my fellow New Yorkers following the advice of medical experts, there is—if I'm being totally honest—an irrational part of me that understands the primal anxiety at the root of those who act out against the loss of the face. (Dogs, I've noticed, are similarly freaked out by covered visages.)

You may protest that our faces were *always* social masks, adapted to the standardized protocols of the general public interface, rather than an unmediated, idiosyncratic index of our true selves. But the fact remains that since we climbed down from the trees, the legibility of our facial features has been the primary mode of "reading" one another's moods, intentions, reactions, and so on. To see only the eyes is to realize how much the gestalt of the face is necessary for this vital kind of legibility. Apparently, we need more than "the windows of the soul" in order to connect. (Muslim cultures have no doubt learned to adapt to this narrowing of the aperture on to the Other and, over the centuries, invented various compensations, sensitivities, and strategies of decoding.)

What a relief it will be, when faces return *en masse*; when they shed this new and sinister role as enabling accomplice to a set of infectious vectors; little more than a fleshy palanquin for lethal and leaky orifices. The return of the face will be a Levinasian re-gifting of something that we took for granted for so long. Perhaps we'll even read them more closely, rather than viewing them as a quick and blunt way of sorting our fellows into general categories. ("That face over there? . . . TL;DR.")

August 15, 2020

Continuing the theme of lost or endangered singularities, I have been fascinated by UNESCO's Intangible Cultural Heritage register since I learned of its existence. At present, this initiative includes such diverse practices and forms of knowledges as Byzantine chants, Ethiopian epiphanies, ritualistic practices of the date palm, Irish harping, and the Celestinian forgiveness celebration. (With the French *baguette* up for a vote soon.) Were I to find myself in the privileged position of overseeing this register, I would lobby to include some of my own ever-growing list of nominations for "intangible heritage," including:

The evening light on Sicilian sandstone churches
Rhyming Cockney slang
The smoky scent of an open fire
The cold side of the pillow
The first gulp of beer on a stifling hot day
Doing "The Robot" (un-ironically)
The sound of ducks landing on water
The sound of a kookaburra laughing (especially when you've just done something foolish)
Phatic, nonlinguistic Japanese conversational noises
Australian mixed lollies
Mom jeans (pre-1998)
The first time you hear Led Zeppelin's *Whole Lotta Love*
Side-boob

The confidence of the French
The earnestness of British Columbians
Spooning
Aged Comté cheese
Tact
The smile of a soon-to-be lover
The way Irish people say the word "film"
Roller discos
Dolly Parton
Well-used copies of *The Moosewood Cookbook*
The smell of a good second-hand bookstore
Winking
Bringing the beat back

Feel free, dear bees, to add your own.

August 16, 2020

Bonjour, mes chères abeilles.

In my dream last night, I was asked to summarize America in two words. Without hesitation, I said, "violence and bacon."

August 20, 2020

Today I woke to find that local blogs and mailing lists are indignant at the news that hotels in the neighborhood, empty of guests because of COVID, are now being repurposed as halfway houses for the homeless or unhomed. While many are upset that (some) mentally unstable people are being bused to our streets—only a tiny percentage of whom have even been convicted of violent crimes or classed as sex offenders—I am more troubled by the Band Aid solutions that the city is applying to deep and chronic problems. Over recent years, more and more mental health facilities, clinics, and even whole hospitals have been closed, purely in order to save some money in the long run or make some unscrupulous huckster richer in the short term. Meanwhile, people who need care and treatment are left to fend for themselves. (I would certainly not mind paying hefty local taxes if they went to social services, rather than directly to the police.) As hundreds of thousands of people flee the city for places like Hudson and Kingston, in a reflex replay of the infamous "white flight" of the 1960s and '70s, headlines announcing "the end of New York" surface online. Certainly, I have been guilty of romanticizing "the real New York" of the gritty old days, even as I understand that any return to such would be a case of "be careful what you wish for." Yesterday, for instance, while buying a MetroCard in the 72nd Street subway station, a woman was stabbed at lunchtime, in broad daylight. There was no apparent motive, not even robbery. Instead, the assailant just wiped the blade on the victim's clothes and kept walking. Reports that the attacker was one of the 200 men who have been put up in local hotels have not been confirmed. Many of my neighbors are understandably jittery. And I can't pretend to be immune, even as I recognize most of this "concern" to be a flagrant case of NIMBYism.

Walking in the Park before COVID hit, I noticed a homeless man walking toward me. He was talking to himself, but not in an agitated way; rather, he seemed to be deeply engrossed in a conversation with a ghost from his past. He was probably in his early forties, bearded (of course), with gentle, haunted eyes, and I don't think he could even see me, since he was reliving some moment that refused to leave him be. "I understand," he said, softly, as he passed by. "I wish you all the best. I really do."

August 23, 2020

I was reading over some notes for a research project and was struck once again by this remarkable insight from Henri Lefebvre (written in the 1980s but becoming more relevant with each passing decade): "Computerized daily life risks assuming a form that certain ideologues find interesting and seductive: the individual atom or family molecule inside a bubble where messages sent and received intersect. Users, who have lost the dignity of citizens now that they figure socially only as parties to services, would thus lose the social itself, and sociability. This would no longer be the existential isolation of the old individualism, but a solitude all the more profound for being overwhelmed by messages."

August 24, 2020

We are heading toward seven months since the reality of the pandemic dawned on the world, and we all beat a hasty retreat into our private dwellings, like frightened and wary hermit crabs. Constitutions, both personal and political, have been tested, and relationships have been challenged ever since, especially intimate ones. (I know several couples already who have not been up to this challenge and have now gone their separate ways.) Many of those who are shackled and shacked up have no doubt glanced with envy at those who live alone, while those who have been enduring extended lockdown in solitude no doubt wish that they had someone with whom to share the ongoing eventlessness of the passing days. The grass is not only greener; it seems farther away than ever.

In my case, as a committed card-carrying Cancerian, staying at home is less a sentence than a default mode. I have been training for this kind of situation for de-cades. And yet, I underestimated the importance of those routines that took me beyond home base—to the office, to the dentist, to the concert. All those tiny social interactions with colleagues in corridors, and strangers in the street—even silent and fleeting ones—add up to the essential texture of life; of *a* life, lived in motion among other lives. Humans are designed to be in at least minimal Brownian motion with one another, like vibrating, mutually mobile molecules. So to say, being trapped inside a 700-square-feet two-room apartment with one's spouse for essentially twenty-four hours a day inevitably starts to grate.

Meanwhile, a close friend—and partner in various creative crimes—has found herself with only herself for company since the first week of March. Thankfully, this friend is very good at squeezing interest or pleasure from even the most banal moments or elements. But even a beautiful house, limitless streaming options, and all sorts of exotic deliveries do not distract from what is essentially solitary confinement. Nothing can compensate for the loss of *touch*, since we are mammals, after all, and mammals need warm and fleshy tactile reinforcement. And so I despair on her behalf, as well as that of all who find themselves in the same position.

Elias Canetti writes, "To live, what you need set out before you—more so than any number of goals—is another *human* face." I would take issue, however, with the word "human," since a dog or a cat can quite often provide deeper and more enduring creaturely comfort than people. But for whatever reason, my friend is hesitant about adopting a pet, even under these extreme and exceptional circumstances. (David Abram asks, "Do we really believe that the human imagination can sustain itself without being startled by other shapes of sentience?") In any case, it's beyond frustrating to all of us to be so close to friends and family but in most cases not being able to see them in person. (At least, not without gravely endangering them and/or oneself.) Anyone farther than a bicycle ride away may as well be living on the moon at present. Nevertheless, this grim fact does not stop me from trying to figure out the logistics of doing so, along the lines of those puzzles from high school tests. (*"You are charged with moving a goat, a monkey, a bale of hay, and a bunch of bananas across a river. How do you accomplish this, with only four journeys from one bank to the other?"*)

To pass the time, I have recently started watching an orphaned baby otter, by the name of Joey, on live cam, rescued by a zoo up in Canada. Somehow, spending time looking in on fellow creatures—themselves trying to cope with bewildering captivity—helps us feel less lonely, and less personally afflicted by this surreal global bardo. The eye, however, must act as the haptic organ, reaching out to touch, stroke, and caress the beloved critter on the other side of the screen. Something about the determined, idiosyncratic spirit of these animals buoys our spirit. And yet the fact that we cannot embrace them—cannot feel them, smell them, hear them— ultimately serves to reinforce and measure the extent of our mediation, and the degrees of our alienation. I also check in on a Russian bear, Mansur, who lives a largely solitary life and is forever trying to improvise games and encounters with assemblages of objects that mimic bear-like contact and interaction. He loves anything that resists and perhaps even pushes back. He loves his giant ball in the pool, or a piece of wood twisted into a ravaged hammock, like a spring. Meanwhile, Joey the baby otter forever play-wrestles in his crib with purple and green blankets, designed to mimic the enfolding fellowship of kelp.

Having a warm-blooded companion certainly takes the existential edge off this infernal limbo, in the sense that one can use the body of the other to confirm that one still exists, in a direct, infantile, tactile way. One still has some kind of bodily purchase on life. Some of my friends, however, have experienced the situation where the fundamental reassurance of presence can tip into suffocation and claustrophobia. (Always a danger, in any scene involving two protagonists.) Being alone, however, is no picnic. And even monks would break their hermitage and gather for meals and ceremonies at regular intervals. On the other hand, the *I Ching* insists that one must learn to dwell in the profound solitude of the self and also learn to travel great distances without leaving one's room.

Which is all to say, my dear bees, that I hope you can take at least some time each day to appreciate the communal hum of your hive. Compressed collectivity no doubt has its challenges, even insectoid ones. But your species has not yet been obliged to invent technologies of solitude, nor solitude itself as a technology of being. (Or, indeed, bee-ing.)

August 26, 2020

Another day, another flagrant execution of a Black person by the police, all caught on video. This time, two protesters were murdered by a seventeen-year-old militia member who drove across state lines with his (legal!) automatic weapon, just to indulge in some human hunting, for sport. Last night I happened to watch the new Brazilian film *Bacurau*, which could be described as *The Hunger Games* for the Criterion set. These cinematic dystopias, however, are less and less "enjoyable" as the world rapidly begins to overtake their own extreme conceits. I suppose I should no longer be shocked by the fact that the police were also filmed aiding and abetting this homicidal cosplayer in his despicable mission, leading up to the kills, while they also murder twelve-year-old kids, like Tamir Rice, for the crime of playing cowboys and Indians in a park while Black. But it still makes the blood run cold, the stomach churn, the mind reel, and the teeth grind down, harder each day.

August 28, 2020

Something about the apocalyptic tenor of the times, combined with the boredom and stress of being cooped up at home, has meant that New Yorkers are flooding into Central Park—and especially into my beloved Ramble—in plague-like numbers. This miraculous forest, in the middle of Manhattan, was notorious in the 1970s and '80s for muggings and *al fresco* gay cruising. More recently it has been a cherished daytime escape for me and my neighbors, and a rather miraculous site for what the Japanese call "forest bathing." In yet another sign that the city is returning to those grittier days, however,

I walked around the winding and leafy paths of the Ramble this evening, amazed at the number of single men, wandering in and out of more wooded areas, clearly on the make. The libidinal traffic is so busy and apparent that it looks like an exaggerated scene from a 1970s British sex-farce, sped up for comic effect. Normally, during my morning walk, I see only the aftermath of these post-work Dionysian assignations: contraceptive wrappers, emptied tubes of lube, and even used condoms, discarded near whichever tree trunk served as support for a sweaty moment of anonymous sodomy. Walking around in the evening, in the midst of what feels like a rather compulsive and joyless Bacchanalia, inspires a rather complex and queasy feeling, hovering somewhere between amusement, anthropological curiosity, and general weariness of the usually clandestine sexual rituals of humans. After all, it's difficult to enjoy a sunset stroll when literally dozens of solitary men are furtively, and not so furtively, loitering and passing by in a forest setting, giving one another signals that only the initiated can read and generally creating an atmosphere of roaming banditry. On two separate occasions, one after the other, two businessmen even exited the busiest part of the thickets entirely shirtless, not even attempting to disguise their post-coital relief. Soon after this, a corpulent man in an expensive-looking wheelchair—reclined almost completely horizontally, so that the vehicle looked like a mobile hospital bed—whizzes past me in the gathering dusk. His face, under a mask hooked up to a portable oxygen tank, has an impatient and leering expression. With a tug at his controller, this sickly—yet driven—cyborg rattles off the main path and down a dirt track toward his goal. Meanwhile, this man's weary attendant (and, I suspect, longtime companion) follows slowly on foot, at a distance close enough to keep tabs on the man's whereabouts but far away enough so as not to be obliged to witness in too much detail the operation awaiting him.

August 29, 2020

I passed a couple sleeping rough under one of the bridges in Central Park this morning. They were probably in their late thirties, huddled under visibly filthy sleeping bags, lying on flattened cardboard boxes that seemed unlikely to provide adequate protection against the damp, clay ground. Their eyes were screwed up against the gray morning light, as much as against my own trespass through their makeshift bedroom. It had rained hard the day before, and this unfortunate couple had clearly been caught in the downpour, as some of their clothes were spread out on rocks nearby. Such a humble gesture seemed to symbolize the residual traces of optimism, in this case hope that one's clothes might dry, thanks the boundless blessing of the sun. But the morning was damp, and the re-gathering clouds were discouraging.

Across the Atlantic, the French government has pledged to continue covering the lost salaries of furloughed employees at 87 percent of their former wages. Six months into the pandemic, and the U.S. government expects its citizens to somehow survive on a paltry $1,200 check that many people—certainly the most needy—never received. Improvised shanty towns are sure to return to Central Park, just as almost every neighborhood in Portland and San Francisco has its own Little Hoovervilles. (Many of these pre-dating the coronavirus, it must be said.) As the affluent continue to stream out of New York City and hoover up real estate along the Hudson Valley, we are likely to see new Seneca Villages spring up in the Ramble, a stone's throw from Park Avenue and Central Park West. The hope among many is that this city will now become affordable again, no longer the playground of the wealthy and their obnoxious offspring but a place where regular folk can make a home for themselves and bring some organic vitality back to this Disneyland for the 1 percent. The accompanying fear is that New York will also return to the days of sudden switchblades, dirty syringes underfoot, and garbage-covered streets. (Even more garbage than usual, that is.)

Encountering this poor forsaken couple this morning puts two stricken faces to a churning sea of misery and neglect. The people of this country have been completely abandoned, by both political parties, not to mention by their fellow man. The remaining bourgeois class, here on the Upper West Side, has managed to pressure the local council enough to force the homeless people out from the local hotels, which had been briefly repurposed as shelters. The French, by rights, should demand repossession of the Statue of Liberty. For here, now, the tired, the poor, the new huddled masses yearning to breathe free, are wheezing through frayed masks and slumped under shopping carts, wedged between rusting scaffolding. They have been deemed disposable—a problem of civic waste control.

Meanwhile, I walk past two singular examples of the vast human cost of abstract policies—or rather, lack of policies, and lack of basic compassion—under my own boutique mask, averting my gaze to give them a semblance of privacy. My own walking—my own continuing on, continuing past—tracing the same arc of irresponsibility, writ large on Capitol Hill.

September 7, 2020

I still can't shake the feeling that everyone else is on campus, holding classes and attending meetings in person. This is not because I'm especially paranoid or worried that the world is going about its business without me. Rather, it is because I can't fully process the loss of actual interaction that the stay-home order has created. It feels like I'm stuck at home because of injury or illness, while my colleagues are surely still in those meeting rooms, classrooms, and offices. It's difficult to fathom the extent to which the public sphere has shrunk and flattened to the size and dimensions of a computer screen.

October 2, 2020

I have been neglectful again, haven't I, dear bees? The new semester has kept me busy, as I try to adapt a familiar course for the new "delivery system" of teaching online. The whole profession—as well as the students themselves, of course—has been wrestling with this sudden virtuality, in which we are "together" on the screen, arranged in what people have started to call Hollywood Squares or Brady Bunch tiles. The experience is certainly what we also like to call "suboptimal," although the timing is fortunate, given that this new crop of students is of the generation whose members are comfortable expressing themselves through screens and via networked media. (Though it's notable that each new crop of students claims to be the last fortunate humans who managed to experience life "before technology.") All the protocols of muting and unmuting, sharing materials via chat, peering voyeuristically into one another's rooms and lives, is novel and thus somewhat interesting. But, of course, it is no substitute for the experience of sharing thoughts and ideas in the same intimate space and time. (Though one should not overstate the "magic" of that experience either, which can just as often feel leaden, uninspired, and futile.)

But telling you the tedious minutiae of my vocation is not what brought me outside today, into the fresh autumnal morning, but rather the breaking news that the Idiot King, the tyrant and bully-in-chief, has tested positive for the virus and has even been hospitalized.

People are celebrating online, even as many—me included—are skeptical, since anything and everything he does is suspect and designed to distract. (This news comes only a day or so after it was revealed that he has been paying a whopping $750 in annual taxes for years and bungling many, if not most, of his dubious "business" ventures.) In any case, this seemed like the kind of news that I'm supposed to relay to you, my pollen-drunk friends, since the whole world is buzzing with speculations, hopes, and fears concerning the possibilities ahead, with the election only a few weeks away.

November 1, 2020

I've been very lax, haven't I, dear bees? This is my second post in two months. Please don't think of this as negligence, or a loss of interest in keeping you abreast of things in the human world. I think of you every day. But things have been so busy, adapting to work online, that I haven't managed to put on my imaginary beekeeping suit and visit you.

Truth be told, I also don't want to depress you. But the news is almost uniformly grim. The virus swirls around the world, cutting down millions, with no vaccine in sight. (Although it spared the Idiot King, for some accursed reason, who seems back to his usual inexplicable picture of reanimated health.) We're all still mostly confined to our houses and apartments in either self-imposed or government-mandated lockdown. The weather in the Northern hemisphere is switching to winter, and along with it the dreaded "second wave" of spiking cases. I'm finding it especially hard, as autumn is usually the most comforting of seasons, sprinkled with folksy customs that aren't my own but that I've happily adopted. Indeed, the hop-skip-jump from Halloween to Thanksgiving to Christmas usually fills me with a glowing sense of impending coziness—as it does for so many—wrapped in pleasant, wood-smoked melancholia. But this year is obviously an exception, when we are all cowering indoors, filled with dread and trepidation.

Last night was Halloween, but with no real trick-or-treating, given the ongoing social distancing. Such a shame, given that it was the first full moon on Halloween since 1944. (Another banner year.) I feel so sorry for the kids who are dressing up, but obliged to wear a mask, though it's not the usual mask one typically thinks of in connection with Halloween. Yes, my heart is fairly bursting, truth be told. For everyone. For all of us.

One of my grad students, K., put an end to his own life a couple of weeks ago. K. was incredibly vital and enthusiastic, and I didn't see it coming at all. (Even though his writing was suffused in masochism and trauma.) This young man transitioned just last year. But even as K. was becoming more himself, and more comfortable in his own skin, he did not want to go on. It's a cliché that happens to be true: Sometimes the people struggling the most are also the best at hiding this fact. K. was so busy looking after everybody else—setting up care networks for LGBT outcasts and organizing workshops for grad students to share their work, helping professors with the logistics of making their ambitious feature films—that he did not manage to care for himself. K's final irreversible act blew a huge hole in dozens and dozens of young lives, especially in the cohort of our students, who were all so close. The memorial—mostly conducted by Zoom (though some gathered on a rooftop in Queens)—was a somber and surreal affair. Seeing all those grief-stricken faces in those infernal thumbnail tiles but not being able to hug anyone, or simply be in the same affective space, was very difficult. In what was a bit like a morbid Quaker meeting, we took turns giving our testimonials and sharing our memories. But there were also long stretches of silence as we just looked at one another, unsure whether there were technical difficulties or just deep lulls in the will to communicate. Nevertheless, I am glad it happened, as there was at least some kind of simulation of a wake, even if, in this particular *annus horribilis*, there is no catharsis or meaning to be found.

Speaking of which, the general election is in a couple of days. I'm sure you bees have felt the agitated buzzing increasing as the date nears. The social networks are a-twitter with an endless flow of enraging and terrifying stories and scenarios. The current administration is not even trying to hide the fact that its members are trying to steal the election: rubber-stamping blatant gerrymandering plans, sabotaging mail-sorting machines, installing fake ballot boxes, and purging millions from voter registration lists. The sense is of a bunch of gangsters, selling off to the highest bidder even the copper-wiring of this formerly aspiring democracy. Wealthy New Yorkers are rumored to be hiring off-duty cops to patrol their gilded apartment complexes with submachine guns. Walmart can't manage to keep guns in stock as state senators talk openly of civil war. Meanwhile, uptown in The Bronx, a man innocently eating a deli sandwich as he walked along the street fell through collapsing pavement into a rat-infested vault. (His friend noted that the man was too afraid to yell out for help, in case rats might run into his mouth.) If this isn't an allegory for the United States right now, I don't know what is.

At any rate, I try to take each day as it comes, even as the eternal return aspect is really starting to carve strange ridges into the psyche. (Not to mention its playing havoc with my neck.)

I did manage to escape the city, for one much-needed week in that magical locale, "elsewhere." We rented a car and drove to the Catskills. Hiking through mossy forests, without any need of a mask, felt like a luxury and a tonic. I bought a homemade blackberry pie from an old lady in a small, picturesque town, who was selling her baked goods from a cooler on her lawn, even as she loitered and waved behind her kitchen window. (No doubt to protect her health.) Payment was cash only, and according to the honor system. The pie itself was delicious and seemed to embody the very opposite of all the trauma and distress of the past few months. Driving back into the city, however, was an apocalyptic experience, suddenly lurching from golden autumn leaves to disheveled streets and homeless people wearing masks over their eyes as they walk, like something out of a zombie movie. The dismal scene reminded me of a moment in Maeterlinck's *The Life of the Bee*, describing the aftermath of "the massacre of the males": a horrifying scene in which the peaceful workers suddenly turn into judges and executioners. "The atmosphere of the city is changed," he writes. "In lieu of the friendly perfume of honey, the acrid odor of poison prevails; thousands of tiny drops glisten at the end of the stings, and diffuse rancor and hatred."

November 2, 2020

Cold and windy today. At least thirty raptors are circling above the buildings. (I've never seen more than two or three hunting together like this.) An old woman in a white nightgown is pacing up and down manically on her balcony, more than twenty stories high. One hopes she's just trying to shake the anxious doldrums out of her weary bones and not trying to psych herself up into jumping.

Meanwhile, I'm still trying to shake off the dream I had last night, in which the sun had risen but everything was still dark.

November 5, 2020

It has been two days since the eligible humans in this country have voted, and the entire world is still holding its breath. (For while the United States is clearly in sharp imperial decline, it still, for some unfathomable reason, holds a disproportional amount of power in the international arena.) Given what we have all been through, few of us dare to be optimistic. And even the best-case scenario, in which the Idiot King is deposed, would signal merely a return to business as usual. No one is sleeping. No one is smiling. Everyone is on edge, as the newspapers speak of possible civil war just around the corner. Indeed, Walmart is storing behind the counter this week what guns they can keep in stock, in an unprecedented move to either minimize violence or at least get some kudos for seeming to do so.

November 8, 2020

Hello again, my patient bees.

They say twenty-four hours is an eternity in politics. And we've certainly been dwelling in a kind of data-saturated purgatory for what seems like forever since the polling stations closed. But as I'm sure you could feel, the humans—at least the ones more numerous in this part of the country—have suddenly been transformed. Those noises you heard—hollering and honking and cheering—were the sounds of long-lost relief and joyful disbelief flooding back into our hearts: a profound shock that this accursed year actually delivered a surprise, happy ending. (At least in theory, as a wounded and cornered tyrant can be very dangerous, in direct proportion to how pathetic he is.)

The sound began as a low murmur, a kind of sonic tide. At first I thought it was a street demonstration approaching. But the mood was light, buoyant—jubilant. Then it grew and grew. As with the most terrifying moments last winter, people were banging pots out of their windows. But this was not in grim solidarity with medical workers but with an infectious glee at seeing the Idiot King finally unable to slip from the grip of the people—like an expertly greased pig—as he has done so many times before. Until this moment, the whole sour ceremony of the election felt like democracy entirely detached from the *demos*. But suddenly the world shifted from sepia to Technicolor. (Indeed, some people were singing out their windows lusty versions of "Ding dong, the witch is dead.") There was spontaneous dancing in the streets, as if the coronavirus itself had suddenly been vanquished, by some miraculous Saint George. We especially cheered the U.S. Postal Service trucks, which formed convoys and beeped their glee to the people they have so faithfully served through snow, rain, heat, and gloom of night.

It never occurred to me I might live in the kind of country that would celebrate the end of a head of state as if it were the end of a long dictatorship. And while I knew the population was traumatized by the whole experience, even before the pandemic, I was not prepared for the somatic relief of this outcome. I suddenly felt fifty pounds lighter. Suddenly my lungs were working in a way that even a disciplined regimen of ancient Chinese exercises could not accomplish.

Doubtless, the old battles will loom up afresh very soon. But for the moment, this is enough. For the present, it feels miraculous to have something to celebrate at all.

Dare we hope, dear bees, that new and heartening surprises wait around the corner, now that we have a taste for them? A taste we now crave, like blood for a vampire!

November 9, 2020

Of course, the jubilation could not last.

Nor even the sense of relief. For the Grifter-in-Chief is refusing to acknowledge the election results, just as he refuses to acknowledge most other landmarks that guide what one of his flunkies has referred to, rather patronizingly, as "the reality-based community." As Brecht so perceptively wrote: "Do not rejoice in his defeat . . . . For though the world has stood up and stopped the bastard, the bitch that bore him is in heat again." In this case, however, the bastard is still with us, holed up in the Oval Office, presumably surrounded by fast food, armpit sweat stains, and qualm-deficient lawyers with dubious credentials, and even more dubious hair-plugs. The Republicans have the gall to accuse the other side of vote-rigging, even as they tried to steal the election in the plain light of day—dismantling postal sorting machines, illegally purging voter rolls, installing fake drop-boxes, and gerrymandering districts until they look like crossword puzzles.

The upshot is that New Yorkers are starting to get that sick and queasy feeling again, after two days of brief relief. We feel like those fools in the horror movie who cheer with elated exhaustion at the slaying of the monster, even as twenty minutes remain in the running time and the monster's claws are twitching in the shadows. Meanwhile, the *invisible* menace infecting the world is still raging through the population, completely out of control. Each day sets a record for positive cases, and we learn that even apparent full recovery can leave a legacy of shredded lungs, compromised cardiovascular systems, depression, and even psychosis.

And so I continue to avoid human interaction beyond this abode, except through the screen. The notion that I technically still have an office downtown is surreal, the books and furniture gathering dust. The idea of meeting a colleague for a coffee is downright fantastical. What it will be like, if we eventually "get back to normal," is hard to wrap one's mind around. Will we weep at the accidental touch of a bodega cat? Will we throw our arms around those people with clipboards soliciting money for charity? . . . Or will we swiftly click back to our habitual urban mode, avoiding eye contact and cursing the return of the tourists, who slow us down by only a fraction of a New York minute?

November 14, 2020

A walk in the Park this morning.

Which is no longer "a walk in the park," as it were, as the pleasures of perambulation are complicated, and even neutralized, by the vigilance needed to spot and avoid other humans. (Especially those dragging in tow a plague vector, now that we know that children are the main "super-spreaders." Or at least, so the latest news tells us.)

In one especially picturesque section of the Ramble—where a little stream bubbles along a tiny ravine, covered in colorful autumn leaves—a large hawk was hunting. Rather than swoop from the sky, as one would expect, it was sitting on low branches and then simply hopping down onto the ground, among the rustling leaves, rather comically running after squirrels. We watched entranced at finding ourselves so close to the hunt, the hawk seemingly oblivious, or merely indifferent, to our presence. One squirrel nearby was hanging on to a tree trunk for dear life, afraid to move and hoping the raptor hadn't spotted him. We could see the squirrel's furry little body hyperventilating and making little noises to warn others, even as he hoped to continue undetected. As nonsquirrels, we found it indeed difficult to tell if he was bravely alerting his brethren to the imminent danger, or if he was merely letting out little yelps of fear; for nature is indeed red in tooth and claw. Thankfully, however, we were not witness to any carnage, and eventually continued on our way, when the hawk decided to try another part of the Park.

A minute or two later we happened upon a different kind of fauna: a plump and friendly-looking homeless man, shirtless and sunning himself on this unseasonal day by a little wooden bridge, like a hairy troll, too amiable to demand coin for our passage. We pulled on our masks upon seeing him, and he announced theatrically, "Ah, what have we here? A couple of bandits." I couldn't be sure if this cheerful soul could see my smile beneath my mask. Though I know he heard my muffled, unimaginative words, "A beautiful day," because he responded, "It is indeed." Such are the random encounters and exchanges that make me still love New York, even as it crumbles back into a kind of forsaken plaguescape. Fleetingly, very different people—perhaps endowed with similar spirits—can see, recognize, and acknowledge one another, among the clamor and the chaos and confusion.

Indeed, for some reason this meeting also made me recall a very old but preternaturally sprightly woman who used to live in the same large co-op building on the Lower East Side, several years ago, as I. She dressed as if she were still living in a nineteenth-century Ukrainian village and was always on the move, restlessly following wherever the mischievous twinkle in her eye led her. And on those occasions when the elevator doors opened, and she was waiting there inside, she would pretend to look scared before announcing in a theatrical manner, in her thick Eastern European accent, "Oh no! Teenagers!" . . . and then mime running away in fear. (At this point it should be noted that we were already well into our forties.)

November 20, 2020

Once again, briefly escaping the two rooms that have become the extent of my entire world and scurrying to the Park for some chilly *shinrin-yoku*. Sitting on a bench, in a discreet sun trap, I hear strange noises from the other side of some landscaped bushes. A new guessing game is thus born: "Kids' entertainer? . . . or deranged person?" Under these conditions, the chances of being right are approximately 50/50. Especially as the paid entertainers are likely accomplished—perhaps even celebrated—musicians, reduced to singing artificially perky children's songs in the Park behind itchy, saliva-soaked masks, to rich toddlers who have no notion of either social distancing or the etiquette of attention. So to say, yesterday's Broadway talent may well be tomorrow's deranged person, thanks to today's gregarious indignities.

After walking a while, to get away from the disturbing noises behind the shrubbery, I notice the ever-present Eckart Tolle guy, who stakes out the same bench nearly every day, overlooking a favored birdwatching spot by the lake, known as The Oven. He's called the Eckart Tolle guy because he has a little shopping trolley full of copies of the bestselling self-help book *The Power of Now*, which advocates the holistic benefits of living in the moment, rather than dwelling in the past, or anxiously anticipating a future that, after all, may never come. If ever there was a time to absorb and enact the lessons of this Zen-like philosophy, it is now. And yet even the Eckhart Tolle guy is looking glum and impatient, one of his legs jittering up and down. Indeed, I wonder how long it had been since he'd even read the book that he has been such a faithful evangelist for these past several years, for as long as I've been visiting this part of the Park.

Wandering home again—my fists clenched in the *wò gù* style of Chinese energy medicine, which temporarily closes the body-mind off from porous external harm, and wending my way between joggers and strollers and deliverymen, like some kind of advanced, manic video-game level—I ponder how long I myself can live in this seemingly eternal Now of the pandemic. Nine months of social and psychic suspension have played havoc with our temporal senses. Sometimes I wonder if the human race is essentially over, and this is now an extended, post-historical purgatory within which to reflect on our modest accomplishments and an almost infinite litany of mistakes. Is this an epilogue, in some sense? Or merely an interval?

November 26, 2020

Thanksgiving.

Given that I am not American, this holiday tradition does not resonate with me in the way it does with so many in this country. Apparently, it is felt deeper in North American bones than Christmas. Which partly explains why so many people are ignoring expert medical advice and traveling across the country to gather in large inter-familial groups. The second wave of COVID cases—already a tidal wave compared with summer's—will soon be gathering into a tsunami. So, I continue to hide inside, like a turtle inside a box inside a cave, mindful of the irony that this holiday itself emerged from the need to distract Americans from the genocidal origins of their nation, as well as the virus that the pilgrims and pioneers brought in their wake—as terrible as it was convenient for the purposes of starting a new "civilization" from scratch.

In terms of a Thanksgiving meal, I thought it would be fun to make halal-street-cart-style chicken and rice, since I'm not a great fan of turkey. But I neglected to read the recipe closely and thus forgot to brine the chicken in time. So, we had roasted cauliflower instead.

My thoughts, however, were with the people "celebrating" this holiday alone, whether by circumstance, or by civic responsibility.

November 29, 2020

Good morning, dear bees.

I'm still addressing you directly, even though I haven't seen any of your kind in the real world for several weeks now, as the weather cools. I used to assume you hibernate for the winter, like so many tiny bears. The esteemed American Beekeeper Federation, however, has put me straight and explained that you spend the cold season vibrating together, in order to keep the colony—and especially the queen—nice and toasty warm, replenishing your energy with the honey collected over the summer. (And even zipping out of the hive for occasional "cleansing flights," since you wisely prefer not to excrete where you eat.)

I suppose you could say that humans try to do the emotional equivalent of your winter humhuddling—buzzing around inside our lonely home-combs, trying to keep one another warm as best we can, even though our hives have been partitioned into a million separating walls. Indeed, here in the human world there's a deepening sense of trepidation that this will be the longest winter any of us have ever experienced. The unexpected euphoria following the election has subsided into the now-familiar low hum of anxiety, combined with the apprehension of sudden dread, just around the corner. Even as the vaccine is slowly being rolled out, the logistics are staggering, and staggered. I won't likely feel that welcome jab for another six months. Of course, nearly half the country—including my maintenance guy, it turns out—is wary of the vaccine, saying they will refuse to get it. "We don't know what's in it," they say, even as they vape any purple substance they can afford and eat generic-brand sausage patties for breakfast.

These days, in the golden age of "alternative facts," the challenges are just as much cultural as economic or material.

December 15, 2020

Walking through the Park this morning—my mask providing some welcome warmth, for a change—I passed a guardian of some sort, with a group of tiny kids. They were all singing "Ring a Ring o' Roses," with their masks on: "A-tishoo, a-tishoo," they sang, in their tiny, reedy voices, "We all fall down!"

December 17, 2020

Woke up to news of a new variant of the coronavirus, 70 percent more infectious than the one we've been dealing with thus far, seemingly emerging from the mucous vectors of the UK. Experts believe that the vaccine should also be able to deal with this new strain, but it's difficult for anyone to say for sure, given how much we still don't know about the way our new visitor will behave from one moment to the next.

Nerve-shredding, the way that any distant light at the end of the tunnel keeps winking away, taking any sense of cautious optimism with it, as if a malicious candle-snuffer is following us around the room, extinguishing any wick and glimmer of hope, no matter how faint.

December 21, 2020

Happy Winter solstice, my dear bees.

In this longest night of the year, may your dreams be bright and warm and full of sticky golden nectar. I picture you all cozy in your hive, bundled up in bunk beds, six to each waxy hexagon, a soft droning snore playing through the empty corridors like an ambient lullaby. May all the toil of the summer pay sweet dividends in these leaner months.

Looking out my own window, to the human hives opposite, I see new tenants taking advantage of the lower rents and occupying the apartments that were so swiftly evacuated a few months earlier. The government somehow still expects the population to be living on the $1,200 paid out more than six months ago, while Congress argues over the pork distribution in a new bill that would provide another single payment of $600 for all U.S. citizens in the New Year. (Even as defense spending to the tune of $750 billion was approved without any fuss whatsoever.)

Staring across at my new neighbors, I see a handful of random members of a species starting to reconcile itself with a new, primarily digital and immobile existence. Isolation is the norm. (And who can say if sharing such isolation with another makes the situation easier or harder to bear?) One new neighbor is a man who appears to be in his late twenties, and who lives in a tiny studio apartment. He has turned his desk to face the window, and he works on his computer twelve hours a day. Before bed, he takes two steps to his left and mounts an elliptical machine, working the levers and pumping the pedals in a kind of exhausted frenzy, like some kind of haunted guinea pig, trapped in the bedroom of a spoiled and sadistic child. It suddenly strikes me that if Hitchcock had made *Rear Window* in 2020, during the lockdown, each apartment would reveal only people typing on computers, working out on virtual reality–enhanced gym equipment, or eating on sofas in front of giant LED screens, while small children clamber over the furniture. Not exactly the visual building blocks of a classic urban thriller.

Once again, the key question nags at me: Are we going to re-embrace and re-invent the possibilities of the public sphere? Or are we going to realize that our deep-seated impasses and issues follow us back outside, and not much has really changed?

December 22, 2020

A week or so ago I tried to buy a Christmas tree from the bodega on my corner, as I have the past few years, but none of the resin-scented employees milling about seemed willing to come to my aid. Eventually a man— who may have been an escaped convict from Central America, with a white, jagged scar running down from his hairline to his chin—quoted me absurd prices for the sad and sickly pine trees leaning against the wall. It felt as if he was actively, and even aggressively, trying to dissuade me from bothering him further, with anything as demeaning as a sale. I tried again on two separate oc- casions, with two different employees, with very similar results, to the extent that I decided the cosmos did not want us to have the cheering compensation of an illuminated tree in our apartment for Christmas this year.

Extremely *on brand*, for 2020.

December 25, 2020

Happy Christmas, dear bees.

I feel rather glum, unable to keep my mind from all the people who are trapped at home, unable to see loved ones and family. Santa still stalks the land. But he's mostly taking orders online and sending masked representatives in his stead, driving UPS and FedEx trucks. My mother sends me photos of her and my sister in Australia, swimming in the ocean, eating pavlova, and playing cards on the veranda. A year ago, that country was on fire, hundreds of homes and countless animal lives rendered efficiently into cinders. But now, Australia seems like paradise, practically untouched by the invisible reaper that is stalking the globe. Will flights ever resume? If so, will they be even vaguely affordable? Will I ever see "home" again?

I think, moreover, of the people spending Christmas all alone, many of whom haven't felt the reassurance of human touch for nearly a year. "The other pandemic" is the name we've started to give the mental health crisis in the wake of lockdown. Indeed, my brain, being essentially masochistic, cannot keep looping some YouTube footage that sliced my heart in two. This video was supposed to be heartwarming, featuring a Russian bear in captivity (different from the one I mentioned earlier) receiving a blue plastic cube toy as a present. Apart from the bright blue object, this bear had been living in a very drab and austere pen, in which only gray and brown presided. The poor bear has no trees, no bushes, no grass—only concrete, cement, wire, rubble. When encountering her new, mute, and immobile companion, the bear shook her head with glee, and gamboled around it like a massively oversized baby goat. The moment that destroyed me, however, was when the emotion became too much for her, and she buckled for a moment, as if overwhelmed. One paw went over her eyes in an uncanny gesture that communicated nothing except a sudden and self-conscious wave of self-pity at being reduced to the kind of creature that would be *this* happy, and *this* grateful, for such a cheap and loveless distraction.

Witnessing moments like this, I truly don't understand how we don't all just implode from the sadness—the fathomless loneliness—permeating all things. It's a Herculean effort, to block out the gnawing at the heart of all creatures, great and small. Indeed, on days like today, I tend to agree with the medieval mystics who were convinced that the fabric of the universe itself is woven from a billowing black bolt of infinite sorrow.

December 28, 2020

To speak millennial for a moment: I fear I've forgotten how to person.

January 1, 2021

Happy New Year, dear bees. I do hope you had a good New Year's Day.

The humans aren't sure what to make of, or do with, themselves, since they could not celebrate as usual, either crammed into Times Square or into some overpriced nightclub. Instead, we drank alone and watched our favorite comfort movies. (In my case, *Flirting with Disaster*.) Ringing in a new year doesn't seem to have the same cheerful festivity it used to have when we are so apprehensive about what this one brings. Certainly, we're ready to say goodbye to 2020. And certainly we're ready to officially drag the Idiot King out of the Oval Office by his spongy ear.

So, here's hoping all will proceed without incident.

January 6, 2021

So much for "without incident."

The transfer of power from the Orange Menace to the Senile Great-Great-Uncle was interrupted by an insurrectionary mob who broke into the Capitol, creating general mayhem in the highest seat of power, in the media, and in the hearts and minds of those watching on, aghast. Most Americans are shocked that a noisy rabble could breach their sacred halls like this, forcing senators to cower in their locked offices in fear of being lynched. (And it appears that some of the rioters brought zip ties for just such a purpose. I worried especially for Representative Alexandria Ocasio-Cortez, whom the rabid right seem to enjoy targeting the most.) The casualty list is unclear, but at least one security guard died in the line of duty.

The most memorable image of the invasion to emerge from the fracas thus far was that of a tall and rather burly man who stood in the main chamber wearing a full fur hat with long Viking horns, sporting a painted face in red, white, and blue, like a redneck reincarnation of *Braveheart*. In this postmodern barbarian we have the new poster child of a kind of grassroots rebellion spawned by constant goading from Fox News, Alex Jones, and other peddlers of conspiracy theories, cut and sold by the foot. This intimidating alpha male stood behind the main lectern and flexed his bare, tattooed chest: a beast seemingly summoned from the deepest fears of readers of *The New Yorker*—uncouth, ignorant, belligerent, and baying for blood. "How could this happen?" is the question on many liberal lips. Though it's also true that we all dreaded something along these lines, given the inflammatory rhetoric of the Idiot King and his minions (and even more so, the strategic silences and dog-whistles) concerning a "stolen election" and the need for "the people" to put things right, 1770s-style.

Despite this incident—somehow farcical and historic at once—the bureaucrats of democracy held firm, and the official winner of the election is now president, albeit by the skin of his (perhaps false) teeth. A collective sigh could be felt, brushing across the land like a steady warm front. Though we're clearly not out of the woods yet.

January 8, 2021

I just woke from a dream in which I was hanging out with The Count from *Sesame Street*. He had a real body, but his usual felt head. He was also a rockabilly, wearing tight denim jeans and a Marlon Brando t-shirt, and smoking a cigarette in that weird way rockabillies do.

January 10, 2021

Terror and tedium, all swirled together.

Sometimes I feel like I've been sentenced to at least a year in a progressive Scandinavian prison: allowed a walk one hour a day, and access to books, TV, and a communal kitchen. But otherwise a prisoner like any other felon, obliged to pay his debt to society. Only in this case, everyone is in the same position.

Other times I'm visited by a keen panic, resembling that feeling when you're watching a film on a plane—involved in the story, and feeling almost cozy, or at least content—but then a little jolt of turbulence makes you suddenly realize: "Help! I'm strapped inside a small metal tube, rushing through the air, hundreds of feet above the Earth!" These days, I feel this way all the time: suspended between a white-knuckled kind of forbearance and a sudden, visceral reckoning with the unbearable heaviness of mortality.

The one-year anniversary of the coming of COVID is on the horizon. I'm impressed by how quickly vaccines seem to have been engineered, seemingly out of thin air and unprecedented technology. While we are concerned about a largely untested injection, most of us are willing to take the risk, if it means we don't have to see the Grim Reaper grinning at us from every window or peeping over every stranger's mask we see. The "Fauci ouchie" is first going to be available only to front-line medical workers and those obliged to deal with the public, which makes perfect sense. It's not clear whether I, a college professor, qualify for this first group. (Especially as classes are still only online.)

A glimmer of light at the end of the tunnel, at least. For if these vaccines are as effective as we hope, then perhaps we'll be dancing in the streets, and smooching one another, like VE Day sailors and nurses, by summer.

January 12, 2021

We're approaching nine months since COVID threw the world off the rails, which seems significant. Babies conceived back in March will be filling the maternity wards, their own bloodstreams filled with strange new apprehensions, beyond the usual ones associated with parenthood. Stress babies. The anxious fruit of the new plague.

Meanwhile, it's the banalities of the pandemic that make up the texture of everyday life, reminding us that we can get used to almost anything, provided it persists long enough to be folded into routine. I'm still walking down and up nine flights of stairs, rather than travel in the tiny hot box of our 1980s elevator. I'm putting on a mask just to put the trash or recycling out. I'm still avoiding joggers like aggressive zombies in a high-stakes video game. And while things aren't quite as intensely surreal as the they were at the beginning of the first lockdowns and curfews, the fact that the coronavirus has changed the world forever can no longer be denied. (How indignant and disbelieving we were, when informed we may have to deal with this microscopic foe for a full two months!) Thankfully we can now order food from restaurants, partly in an attempt to keep favorite businesses alive, now that we know this is mostly an airborne virus. We don't need to wipe down all our groceries with disinfectant, as we were all diligently doing for several months. And speaking of groceries, we are thankfully no longer confronted with zero delivery options, and wondering if we'll have to learn how to hunt for and eat squirrel for dinner.

But still, we shy away from our fellow men and women and wonder if we'll ever be able to hug others again.

February 14, 2021

Good morning, dear bees. And happy Valentine's Day.

I was recently reminded that this holiday was originally installed in the calendar to honor the martyr Valentinus of Terni, patron saint of beekeepers and epilepsy. I hope this means you all get a day off from your labors so you can take your sweethearts to a romantic spot and sip honeyed cocktails together. (Or perhaps you're tired of the taste of honey and would relish something more savory.)

Given the number of lovers who have been thwarted by the virus, this is an especially trying day. Normally, this is one of the busiest days for restaurants, florists, and so forth. But the eateries remain closed, except for takeout, and couples must find other places in which to express their affection, often from a distance and through a screen. (Like everything else these days.) My heart is heavy for everyone craving either general affection or a specific presence that has been denied them these past ten months: an eternity, when it comes to the heart, the soul, the skin.

I think back to the first few days of lockdown: an unthinkable kind of incarceration for millions of people in New York City. I was awoken at 5:00 A.M. by a couple of middle-aged men, slow-dancing together on the roof-deck opposite my bedroom window, foreheads pressed together, and weeping openly and unashamedly. Perhaps they knew what lay ahead. Or perhaps they were merely surrendering emotionally to the uncertainty that had clasped us all by the throat. (And obviously members of the gay community are not strangers to the effects—and after-effects—of a devastating virus. They are the veterans we should be looking to and learning from, in terms of how to survive the shock and fog of war against an invisible foe.)

February 16, 2021

Much of the last week or so has been spent online, trying to secure a vaccination appointment. These are harder to get than FreshDirect delivery slots during the beginning of the pandemic. There are a couple of online portals to register, and once your details are in the database and your qualifications verified—i.e., over sixty-five and/or a worker with the public and/or immunocompromised—then you are theoretically able to access one of the inoculation sites set up around the city. Various Twitter feeds provide more than rumors concerning availability, but you have to have a trigger finger with the mouse, because it's like trying to get tickets to see a new supergroup featuring Jimi Hendrix, Janis Joplin, and Kurt Cobain, all back from the dead and ready to rock. Slots vanish in front of your very eyes.

I suppose if I could go out to the depths of Queens, or any of the other boroughs, my chances would increase. But since—like most people—I'm not willing to risk getting into a taxi, or on the subway, I can try only for places within a few miles that I can walk to.

So, my quest for ostensible immunity continues . . . .

February 17, 2021

I got one!

I managed to land an appointment at a government vaccination site—a commandeered public high school—a good hour's walk away on 135th Street in Harlem. I believe I'll be getting the Moderna vaccine, though it isn't entirely clear from the automated confirmation e-mail. I can already feel myself unclenching a little bit. I've been dreaming about this moment for many months and am excited about the possibilities of an inoculated world.

Wish me luck, dear bees!

February 26, 2021

Well, that was quite an experience.

My first adventure for nearly a year, beyond the deeply trodden personal sheep-trail between my front door and Central Park.

My appointment was for 11:00 A.M., but I was instructed to arrive at the facility a half-hour early. The weather forecast was thankfully clear, so I did not have to contend with an hour's walk, each way, in the rain. I put on my mask—with a second one in my pocket, to double-mask inside the building—and headed out into the street, with all the grim determination and self-bestowed mythical import of Odysseus or Aragorn. A soundtrack of nostalgic grunge-metal (Filter, Tool,

Kyuss, Helmet, and so on) helped make the trek feel even more epic. Indeed, I would be lying if I denied that my eyes misted up at several moments, glimpsing the possibility of a post-COVID life—up there on the northern horizon, at the top of Amsterdam Avenue—like a warm will-o'-the-wisp illuminating the cold, gray, filthy, and forlorn streets of Manhattan in midwinter.

When I approached the school itself, I realized that this was more of a large processing station than a boutique clinic situation. The line was already looped around the corner of the building, 200 people, at least, ahead of me. As we waited, we all experimented with the new experience of forming an orderly queue while also trying to respect the new rules of social distancing. Some tried to exchange pleasantries, or ritual complaints, through their masks. Most, however, did not want to engage, for fear of being infected just before being granted salvation (in this case, by the gods of faceless pharmaceutical companies).

The line moved mercifully quickly, which was lucky, because my legs were starting to feel weary. Just after an ID check and entering the building, I donned my second mask. After a series of corridors, I was ushered inside a classic, under-funded inner city school gymnasium. About thirty inoculation stations were set up in the middle of the basketball court, staffed by (hopefully medically qualified) volunteers. To the side was a designated waiting area, where the recently jabbed were instructed to wait fifteen minutes, to ensure that no violent reaction was afoot. Before I knew it, I was waved toward a specific seat by a woman wearing goggles and gesturing with a table-tennis paddle, as if she were directing planes on the tarmac at JFK. I settled into a plastic chair, grateful to sit down after two hours on my feet, and rolled up my sleeve. The person wielding the needle was an older Black woman with fake multicolored fingernails more than an inch long. This latter detail did not inspire me with confidence, as she struggled with tearing open the little paper packet containing each dose.

"This is Moderna?" I asked, to confirm.

"That's right," the woman replied, tapping the hypodermic needle.

"And you got yours recently?" I pried, looking away.

"Oh no," she said, as the sting hit my skin. "We have no idea what's in this, and what it will do to everyone."

"I suppose not," I replied, not exactly reassured. (Especially as I presumed all personnel providing the vaccine had themselves been vaccinated.)

After gingerly rolling down my sleeve, I moved to the waiting area for my fifteen minutes of rest and recovery. A young woman approached me with an iPad to schedule the follow-up shot in a month's time. (Moderna, like Pfizer, requires two doses, a few weeks apart.) It was only after she moved on to the next person that I could really take a minute, exhale, and reflect on what was happening around me, on what I was a part of. Despite all the logistical challenges, dozens upon dozens of volunteers had assembled a reasonably efficient inoculation machine, here in the heart of Harlem. (And certainly, walking from West 80th Street to West 135th was a lesson and reminder in the political economics, and ongoing legacies, of racially structured urban life.) Thousands of citizens were cycling through this place every day, all of us traumatized by the past year, and all of us hopeful of outliving and outwitting the virus. There was a spirit of solidarity and purpose in the air, as I imagine there is during wartime: an overall sense of pragmatic action and human comradery—like Britain during the Blitz, but with a New York edge and an American swagger. I had spent so much time hiding away from my fellow city-folk—and actively avoiding them when outside—that I felt a sudden rush of love for the collective pulse of all those abstract souls who had drawn me to New York in the first place: the magic of true diversity; the miracle of cooperation (often begrudging, but also surprisingly generous) that keeps this experiment in human compression going. What's more, this was starting to happen all around the country—all over the world. I felt that a darkness had been stalking the land, and soon enough, a new, brighter, more compassionate life would surely emerge from this difficult chrysalis.

I walked home rather dazed and exhausted, the adrenaline starting to wear off, but still grateful I was still here—along with my loved ones—to be vaccinated in the first place. While I knew that I wouldn't breathe easily until a week or so after the second shot—nor would I consider catching the elevator until then—I already felt less fearful of approaching bodies on the street. (Though I still flinched when I heard a cough or sneeze.) When I got home, I collapsed on the sofa, fatigued and anxious and cautiously optimistic.

Perhaps the spring will usher in a whole new global golden age, as we pick up the pieces, and hopefully reboot proceedings in a more heedful, and less selfish, mode.

That's the hope I'm allowing myself to feel today, in any case, as a sense of progress finally begins to dislodge the general state of stasis that has been haunting us all.

February 27, 2021

No visit from me today, dear bees. Feeling the effects.

March 1, 2021

Well, that was quite something. My body certainly went into emergency mode and started to battle the phantom infection that the needle introduced into my system a few days ago. I felt like I was coming down with the flu: aching muscles and extreme fatigue, mostly. The sign of my immune system's redirecting resources from business as usual to that tiny army of white blood cells and other antibodies. The technology behind these vaccines is almost untested. The makers certainly rushed the trials, and the FDA has granted only "emergency" approval to turn whoever is willing to get a shot into a guinea pig. Most people, however, seem willing to take the risk, at least in the circles I have access to. "The data" is ostensibly enough to warrant use on humans, even though nobody can say with confidence what the effects might be further down the line. (Everything from blood clots to zombification, according to different parts of the Internet.) Most medical professionals seem to harmonize around a common message: *The house is on fire. Don't pause to argue hypotheticals. Do what you can to survive.*

As a card-carrying hypochondriac who is perpetually astonished by the fact that such a complex system as the human body (specifically *my* human body) can make it to the end of the day without some kind of major malfunction, I believe the idea of an experimental inoculation does make the blood run cold, somewhat. But when it comes to risk ratio, I would prefer to take my chances and "trust the science." (Even as history provides many cases both for and against such faith.) At any rate, it feels good to have made a significant step on the road to a more immunized life, to be able to venture outside again and not flinch whenever a jogger jaunts by, or when a group of kids on the street yell breathily near my ear. Life without masks seems like a luxury—one I'm willing to pay for. Even, apparently, if that price is turning eventually into some kind of unfortunate monster from a David Cronenberg film.

March 29, 2021

Since my first dose of Moderna, vaccines are now available through most pharmacies, and appointments can be booked online. I found a slot near Columbus Circle for my follow-up and was jabbed by a paramedic in the corner of a basement Rite Aid, behind a temporary partition. The woman who jabbed me looked exhausted and was too tired to respond when I thanked her profusely for her service. Nevertheless, I hope that when she can finally catch up on some sleep, she can feel the glow from helping hundreds—perhaps thousands—of people stay safe.

March 31, 2021

Same story as before, in terms of side effects. Strong sense of coming down with something serious, only to feel practically 100 percent once again, thirty-six hours later. Now the countdown begins to *full immunity*.

April 8, 2021

Today is the first day that really feels like spring. Another winter survived, though certainly the most nerve-wracking in my own experience. Walking through the Park, near Bethesda Terrace, we came across a big-budget film shoot, for something set in the days of Edith Wharton. Dozens of extras were wearing extremely detailed *belle époque* costumes. Between takes, despite being outside, they all donned masks and waited in a subdued manner. A logistical underling with a walkie-talkie waved us through so we could walk among all the actors, some wearing top hats, others sporting tweed jodhpurs and leaning on penny-farthings. Given the temporal vertigo of the past year, the effect was quite surreal—as if we had stumbled down a time-slip and woken up in 1890.

At any rate, this unexpected scene reminded me the extent to which everything in modern life may grind to a halt, with the essential exception of The Spectacle. For we can apparently live for extended periods of time without everything except bread and simulated circuses.

April 14, 2021

Full immunity!

Or so they say.

Of course, I take every statement with more than a pinch of salt these days, given all the voices and unknowns involved. But it still feels like a profound relief. And I find myself exhaling with my entire body, as much as with just my lungs. (Which will perhaps never again do the first part of their organic job—inhaling—without at least a tinge of concern and restriction.)

April 20, 2021

To celebrate my new status as an officially inoculated member of the human herd, I decided to visit a close friend who lives deep in western Pennsylvania, the same one who rode out the entire pandemic on her own. (Adopting an exceptionally adorable kitten halfway through, for some belated, extremely rambunctious, company.) Riding Amtrak for the first time since the Before Times was a quiet and cautious affair. Almost all passengers were masked, as legally obliged. I wouldn't say I was breathing easy, but I certainly felt emboldened, thanks to the patented technologies coursing through my veins. The attendant in the café car told me that she was happy to be see passengers again, since, for a year, she and her colleagues had found themselves staffing a ghost train, shuttling across the plains for no clear purpose. "We felt like luggage," she said, "just going back and forth."

It felt nothing short of miraculous to be venturing out in the world again. However, it must have been surreal for my friend to suddenly have another human being in the house, after all this time. Not one to be caught unprepared, she had turned her large basement into a "doom bodega." It was an impressive sight to behold. Enough cans, water bottles, dried goods, and other staples to ride out a multi-year apocalypse for sure. The strength of mind, heart, and character to go through something that distressing, without someone else close at hand to share the anxiety—to help with the mammalian reassurance of touch—is something I hope I never have to summon up for myself. As it is, I'm so proud of her—and of everyone else—who endured the past twelve months without any kind of domestic support.

June 1, 2021

As the weather warms, I find myself becoming increasingly annoyed that the masses have discovered the Ramble. (I'm sure you're not thrilled by the commotion either, dear bees.) As you well know, it used to be one of New York's best-kept secrets, mostly for locals. But now—no thanks to COVID, Instagram, and Google Maps—everyone from around the world has discovered this magical forest in the middle of Manhattan, impinging on the local fauna (turtles, squirrels, blue jays, rats) and the indigenous population (Boomer birdwatchers and gay men cruising for an anonymous quickie). Moreover, thanks to the ongoing desire to socialize outdoors as much as possible, new, opportunistic businesses have sprung up, organizing picnics for special occasions hosted by affluent Upper East Side types in the Park. These can be extremely lavish, with fluffy pillows and fancy table settings, and billowing tent-like structures. Only New York City could dream up something as ludicrous as competitive picnicking!

June 9, 2021

A miracle . . . venturing all the way *downtown*. To have dinner with a friend. (My first time "socializing" in this sense, since the lockdown) J. and I sat in one of these new wooden shacks attached to restaurants and cafés that have sprung up everywhere and tried to remember how to simply be convivial. War stories from the past year were shared, along with the appetizers. It felt good to be talking to a friend again, without the fear of buffering, or one of us being stranded on mute. But I also felt a niggling sense of residual agoraphobia. Indeed, we were both ready to head home, once the sun had set, as if we both half-feared that zombies would start to roam the streets after dark. I headed down into the subway for the first time since COVID had turned that realm into a haunted ghost town of its own. Even though it was only 8:30 P.M., I saw only one other person on the platform: a homeless figure hunched on a wooden seat—unheard of for Broadway–Lafayette, at peak dining time.

When the train arrived, I watched empty car after empty car pass by. "Dammit," I thought, "this subway must be out of order." But when the train stopped, the doors opened, and the conductor announced that this was indeed the B train, with the next stop being Washington Square, I realized it was a regular ride. After hesitating several times, I stepped onboard—not only the only person in this car but seemingly the sole person on this entire train. For some reason "we" crawled between stations, often stopping completely in the dark. This can unnerve me and my claustrophobia at the best of times, when surrounded by other commuters, let alone at night, during an unprecedented pandemic, and completely alone in a flickering subway car. After stopping at a couple more stations, and seeing not another soul, I stepped off long before my destination and high-tailed it faster than Orpheus back up to the street, where I flagged down a taxi. Too soon, I thought. Too soon to return to the underworld.

June 14, 2021

News outlets are starting to talk about "breakthrough" infections—that is, fully vaccinated people still coming down with COVID. This is demoralizing, as we were essentially promised bulletproof status from these new vaccines. Then again, I never fully believed these claims, just as the flu shot is never really a guarantee against catching the flu. Experts maintain, however, that while vaxxed people may still become infected, their chances of ending up in the hospital are now vanishingly small. This is, I suppose, the most important thing. In any case, it's becoming increasingly clear that the coronavirus is something we will have to get used to, like the regular flu, and the Supreme Court, and that it's not simply going to disappear, like SARS. The pandemic will, at some point, morph into an endemic situation, and we will have to learn to live with it, as we do with so many other things that are essentially trying to kill us. (Again, like the current Supreme Court.)

June 25, 2021

My birthday today. The big 5-0.

Were I a cricketer, I would lift my bat and acknowledge the genteel applause for this plucky half-century. Though as any good batsman knows, one should reserve the real celebrations for the full century. Now is not the time to get complacent, or too self-satisfied. Roll up one's sleeves, take a few breaths, and double down on making it all the way.

Originally, before all the troubles, I hoped to celebrate in London, where I was born, in some kind of narcissistic geo-symbolic loop back to my own origins. But obviously that is no longer an option, with planes only just starting to return to the sky, and many bureaucratic hoops still standing between one country and another.

As it is, I've been hearing a strange noise the past few days: a kind of stifled, gurgling cry. I presumed it was a troubled child, calling out a window; or a lonely person in a nearby building, sharing their woe. But I discovered this morning that the sound belongs to a raven: an impressive bird I haven't seen around the city before. How strange that it would show up for the first time on my birthday. Not only that, but it perched directly opposite my window, puffed up its glossy black ruff, and gurgled its strange, rather strangulated, screech directly at me! I stood there with my glass of iced matcha and tried to pry for more information. "What do you want, strange corvid?" I asked through the open window. "What message do you have for me, on this of all days?" The raven cawed at me a few more times and then took wing. An hour or so later, I recalled that

Edgar Allan Poe lived a couple of blocks north of here and wrote his famous poem while living on the Upper West Side. Perhaps he was even inspired by an ancestor of this cranky and portentous visitor.

Certainly, I am not completely immune to superstition, or ominous omens. But I was also determined to make the most of the day. After all, I feel very lucky to be alive, and breathing without the aid of a tube. So once again, we headed downtown and were this time greeted with more general activity than before. Clearly the vaccination effort has made even further strides, and the people are returning to the streets, and even below them. The streets were abuzz with New Yorkers keen to get back into the regular groove: eating, drinking, strolling, coveting, envying. While this was not the European vacation I had planned, it felt life-affirming—surrounded by thousands of people who had all shared the same fears and demons these past fifteen months and who were now collectively returning to a tentative notion of the multitude. Hurled into our own private bunkers, and personal bodies—forbidden to kiss, caress, and even converse—we were now re-learning to interact, joke, flirt, and so on. Like a coral reef after the pyjama sharks have left, we were all starting to poke out our heads and tentacles. It was thrilling.

After dinner we walked to Washington Square Park, which the local news had described as a new unofficial outdoor bazaar-*cum*-festival, after dark. (Much to the chagrin to those who live nearby.) Sure enough, there was a Saturnalian spirit in the air, and the park was thick with abundant humanity: buskers, skaters, singers, dancers, revelers. These were the survivors—grateful and boisterous—grabbing back with both hands the public life that had been stolen from them by a minuscule, invisible thief, this past long year. It felt positively medieval.

We even caught the subway home, and this time the carriage was nearly half-full. Perhaps some semblance of normality really is just around the corner.

September 1, 2021

Today the heavens opened. So much so that our ceiling began leaking over the bed, in the middle of the night, necessitating some quick redecorating. But this was mild compared with what much of the rest of the city was forced to contend with. Truly epic flooding closed the subway, stranding commuters. Dozens of people lost their lives in the wider tri-state area. (Many were trapped in illegal basement housing.) Two images in particular seemed to capture the chaos, both circulating around social media in viral videos. The first was a sodden rat floating in the water, spinning round and around, as if practicing for a synchronized swimming routine; in the other, a man floating on an air mattress in a dirty alleyway, with water several feet higher than street level. He appears relaxed, however, smoking a hookah and watching the end times unfold around him.

September 11, 2021

Twentieth anniversary of that terrible day. Uncannily similar weather.

The memorials this morning seem to be very muted, considering the calendrical significance. A combination of COVID-fatigue and the inevitable passage of time, no doubt. My current students were almost all born after 9/11, so its memory is secondhand and highly mediated. Those who were there, however, feel as if it was only yesterday. As always, we live in different times, and different temporalities. The year, the zeitgeist, the times: These are not a unified experience but a cacophony of competing speeds and clashing mnemonic shards.

As fate would have it, I had a weird dream last night in which I could walk among the clouds. They were almost fake, like a cross between Styrofoam and cotton wool. The whole effect (and affect) was peaceful and slow and—well—dreamy. Eventually, in my aerial promenade, I came upon a large jet plane that had crashed or disappeared at some point. It was just slowly bumping around the clouds like an abandoned boat, nudging against some padded jetties . . . a kind of ghost plane.

December 1, 2021

Now the news is reporting a new COVID variant, called Omicron, seemingly originating in South Africa and said to be much more infectious than the original strain. As a result, I'm starting to get that now-familiar chilling feeling. Perhaps the freedom forged by vaccination is already being reined in, the curtains being pulled abruptly shut. Here we go again, I can't help but think. Buckle up, Bronco.

December 26, 2021

Happy belated Christmas, dear bees. Or Hanukkah. Or Kwanzaa. Or whichever holiday you prefer to celebrate. (Obviously you have your own occasions on which to gather and waggle, so I'm merely offering an interspecies good cheer here and not attempting to foist my own human-centric cultural punctuations upon you.)

This was our second year without a tree, partly because the guys outside the nearest bodega who sell pines to the city folk are still price gouging, and partly because it seems a bit like tempting fate, to return to "full normality" with new variants of the virus swirling around and fatality figures climbing once again. In other words, not really a Christmas-y vibe this year, as before. Perhaps next time.

I have my booster shot scheduled for tomorrow. Another jab in the arm. Another round of side effects. I do hope we aren't now caught in a jab cycle in which we're expected to get vaccinated every six months, even as each new variant will forever be at least a few months ahead of the most recent iteration of Moderna or Pfizer (or Nestlé or Haribo or whoever else decides to muscle in on the action).

January 1, 2022

A new year.

A new leaf.

I hope, dear bees, you are curled up and drowsily daydreaming of all the pollen-jostling to come.

As I'm sure you can feel from within the combs, the weather is unseasonably mild, although we haven't seen the sun for several days. Like every New Year's Eve for years, I'm asleep before midnight, though woken by fireworks and cheers. This year—like the last one—everything is dampened by the pandemic; by the latest variant; by the latest surge of infections, despite the vaccines. My arm is still sore from the third shot (or fourth, if you count the regular flu vaccine) from a few days ago. I got my booster in a large and nearly deserted chain pharmacy behind Lincoln Center. The pharmacist seemed harassed and distracted, with his mask below his nose. Walking home after the shot, I passed hundreds of people lining up on the street for free COVID tests, delivered out of government vans parked by the curb. Some experts estimate that one in four people has the virus in New York City right now.

And so, the walk through Central Park feels once again more perilous than relaxing, or edifying. I can't seem to bring myself to walk up the nine flights of stairs again, however. At some point, convenience eventually defeats caution.

January 8, 2022

The first significant snow of the season. Heading toward the Park, I passed a family having a snowball fight on the street. The little girl—who couldn't have been more than five—caught my eye with hers. I could see her nascent brain-gears whirling, as she calculated the risk–reward ratio. And then, after only that moment of hesitation, she threw her snowball right into my face. I felt blessed.

January 9, 2022

In this extended limbo, everything feels both frozen and rushing at the same time. We are all trapped in Groundhog Day. And yet, the calendar pages also fly off the wall, as in a B-movie montage. Being chronically suspended in low-key dread plays havoc with our bodies, as the fight-or-flight mechanism is flicked on and off, like a light switch being toggled by a bored teenager. In my case, seemingly random moments from my life loop slowly around my mind in a never-ending highlight reel, spliced with random B-roll of places and people from my past. When the body-and-soul is not propelling *itself* forward through time—but rather being held fast by the thumb of "bad infinity" (i.e., just one damn thing after another)—there is no sense of momentum, or narrative arc. Life feels tangibly to be over, in some definitive sense. And yet we still (for the moment at least) *live*. It is a practice purgatory. So instead of plans and projects, the mind—or at least my mind—is filled with involuntary memories (sights, sounds, smells, etc.) of such frequency that they are bleached or leeched of any Proustian significance, or even pleasure. The mnemonic grooves cut deeper and deeper. But this also means the record of my personal experience is being worn out. Or, switching media metaphors, I'm forced to watch the movie of my own life, over and over again, including the daily rushes, the bloopers, and the parts that ended up on the cutting room floor. And I was bored.

Indeed, this is one of the ironies of diaries or memoirs or autobiographies, is it not? That only people who lead rather dull lives have the time to record them. (With a few notable exceptions, of course: Casanova, T. E. Lawrence, Antoine de Saint-Exupéry, Jewel, and so on.)

January 10, 2022

This weird temporality also affects my relationship to the archive of TV and movies. Since we are still trapped here—in Plato's cave, with a digital projector—I can watch pretty much every TV show or film ever made. But in a strange paradox, the more choice we have, the less enjoyable any such decision becomes. I tell myself

I should forget the new releases and dive back into the classics, the canon, even of pop culture. But if I, say, decide to watch *Mad Men* again, then this would be mediated through the dual prism of a doubled nostalgia: first, the general social nostalgia for the 1960s, which the show embodied, and then the layer of my own personal nostalgia for circa 2008, when this "prestige" TV show expressed the zeitgeist of a post-millennial, pre-COVID global mood. I'm an easy mark for nostalgia, it must be said. But even I am aware that it's not something one should encourage or wallow in.

January 11, 2022

Today I came across a line in a novella I'm reading that struck a chord: "Time was merely the mask that eternity had put on to seduce the young." It is rumored that César Aira writes all his books in one sitting, without significant revision. If this is true—and he recorded this observation in the midst of his literary riffing—then he really does deserve his glowing reputation.

January 19, 2022

In many ways, I'm trapped in paradise. I have access to conversation, companionship, touch. I can have a warm bath whenever I want. (Or at least, I can when the water isn't coming out brown and soupy, which it does more often than one might hope.) I can order fancy food through the machine, and it is left at my apartment door by an invisible genie. It's like a very comfortable prison. But we all know that humans thrive on novelty, challenge, change. This stasis or limbo reminds me that time and space are not separate entities but part of a continuum. One must march stridently into the future—physically, literally—in order to propel time forward. One must actually put on one's coat and boots in order to meet, or even make, the morrow. Put one's shoulder into it. In other words, I crave new horizons, and the journey to find them. A holiday is as good as a change.

January 25, 2022

There is a starling that returns almost daily to the small and austere steel cross that sits right outside my living room window, almost completely lost among the industrial fans; ugly (and ineffectual), rusty sound-barriers; and crumbling water towers. This unassuming and utilitarian crucifix marks the top of the religious facility opposite my apartment, and seems so half-hearted that I wonder if it's merely a semiotic concession to New York's civic tax code. In any case, this starling does not seem put off by the anti-ornamental nature of this cross and perches on the top, lifting its shimmering little head back in order to sing its avian song over the jumbled rooftops of Manhattan's Upper West Side. The starling is small even for its own species and cannot be heard among even the relatively calm Sunday hum of the concrete island. But what it lacks in sonic reach, it makes up for in consistency, persistence, and feathery devotion, singing its quiet hymn day after day, week after week, month after month. The building itself seems to be some kind of multi-faith operation, as it welcomes bar mitzvahs as much as Hindu weddings. (Which may explain the demure scale of the crucifix on the roof.) But I am certain that none of the employees, visitors, students, or worshipers can match the purity of this unchristened Christian starling, chirping its faith in a blessed and sun-filled tomorrow with a song unquestionably as seductive as its glistening, rainbow-stained plumage.

February 16, 2022

I arrived in New York in almost allegorical fashion—with a single suitcase and only a few hundred dollars in the bank. I had left a tenured job in Amsterdam, for various reasons, but mostly because my employers were working me like a donkey and paying me like a donkey too. My bank balance was in fact going down each month, since there was no support for foreigners, who—by virtue of their belated status—have no foot in the affordable rent system. So, I jumped across the Atlantic with little more than a hunch and a hope that New York City might treat me better, even though I didn't have a job waiting for me. (Leaving a tenured job in academia for anything other than a better tenured position is enough to get you forcibly institutionalized in most jurisdictions, let alone leaving for nothing at all but a change of scene and a clichéd dream.)

In any case, I arrived not long before Christo and Jeanne-Claude's famous installation *The Gates* in Central Park, in 2005—not yet four years after the September 11 attacks. This public artwork was simple yet effective: thousands of saffron-colored "gates" strewn throughout the Park, bringing much-needed fluttering color to the midwinter landscape. Narcissist that I am, I could not help but interpret these rather ceremonial gates as a welcoming gesture from a surprisingly hospitable metropolis, known also for its capricious hostility. Moreover, walking through these guiding structures offered a royal reception for anyone and everyone who made the pilgrimage to the Park. (At least for those who approached the experience in the same spirit in which the artists offered it.) Even though much fun was made of the project by skeptical locals, it was also a human-scale democratic spectacle. (Indeed, I wonder, dear bees, if your ancestors remember this moment. Perhaps, in their bleary wintry state they mistook the fluttering orange flags for giant, unseasonal flowers.) Nearly twenty years later, I realize that "the gates" could have had a secret meaning, known only to the creators. For while these temporary structures were ostensibly inspired by the impressive *torii* gates, found standing sentinel in front of important Japanese Shinto shrines, they could also have a symbolic link to the kind of gates that mark out the course for a downhill skier. (That is, those springy flags that skillful specialists of the slalom, with especially robust knees, try to navigate around as they hurtle down the snowy mountainside—often catching a slap to the crotch in the process.)

This connection occurred to me while watching the mesmerizing documentary of the Sapporo Winter Olympics in 1972, directed by Masahiro Shinoda. In one especially memorable moment, the film focuses on an idiosyncratic genius: a man by the name of Gaston Perrot, who was employed by the Olympic committee to design the slalom course by setting hundreds of "gates" at carefully calibrated intervals. As the film's narrator explains, "The layout of the gates decides the character of the slalom, so the setter positions the poles carefully. The layout of the gates are like questions that should bring out the best techniques without disturbing the rhythm of the skiing. The setter prepares a tricky exam. He is also like a musical composer. The red and blue flags are musical notes on the snow—a musical score composed by the setter. It is a score which the skiers must play." There follows a remarkable montage of different competitors, from all over the world, walking the course slowly, often scrunching up their eyes for a while—and moving their hands like a gravity-obsessed composer—in an effort to memorize the "melody" and tempo of the different spaces between gates. With expert concentration, they anticipate each change of note, key, and rhythm, imagining themselves into the near future, when they will be racing through this vertical composition at terrifying speed, when they themselves will become a human throughline, threading together a silent concert to deafening applause.

Perhaps I'm giving Christo and Jeanne-Claude too much credit in thinking that they may have had similar designs, as it were, in placing their own gates at intentional intervals, to create a kind of silent symphony in the middle of New York City. In any case, it was an auspicious way to begin this new chapter of my life; hurtling toward new horizons at a deceptively sedate pace.

February 20, 2022

I have been reading a popular new book about mushrooms, written by a biologist with strong literary leanings blessed with a vaudevillian name, Merlin Sheldrake. Buried deep in the pages, this little behavioral nugget is offered to the reader in passing: "Male orchid bees collect scents from the world and amass them into a cocktail that they use to court females. They are perfume makers. Mating takes seconds, but gathering and blending their scents takes their entire adult lives." Why did you not tell me this, dear bees? I can't believe I've been sharing so much with your kind and not being alerted to this charming aspect of the being of bees. (Of course, I also respect your wishes to keep certain arts from the human world, as well as your aversion to unnecessary boasting.) At any rate, what a remarkable endeavor your exotic cousins are engaged in. I can picture them now, assembling a waxen library of heady scents, like industrious apian mixologists; swapping chemical "notes" with friends and hiding secret olfactory alloys from the evolutionary competition.

February 22, 2022

Having switched from Sheldrake to Kierkegaard's journals, I stumbled upon this striking moment of reflection earlier this morning from the genius of Copenhagen:

> Why I really cannot say I definitely enjoy nature is that I can't get it into my mind what in nature I enjoy. On the other hand, I can grasp a work of art; there I can find—if I may so put it—that Archimedean point, and once I've found that, everything easily becomes clear to me. I can then follow this one big idea and see how all the details serve to throw light on it. I see the author's whole individuality like the ocean in which every detail is reflected. He is a kindred mind, no doubt far superior yet with the same limitation. The works of the Deity are too large for me; I am obliged to lose myself in the details. That's also why the expressions people use in their observation of nature are so vapid—"How glorious," "grand," and so on; being far too anthropomorphic, they stop at the outside, they are incapable of expressing the inside, the depth. What to me also seems most striking in this respect is how the great poet geniuses (an Ossian, a Homer) are presented as blind. Naturally

it doesn't matter to me whether they really were blind; the point is that people have imagined them so, as if to indicate that what they saw when they sang of the beauty of nature appeared not to the external eye but to an inner intuition. How remarkable that one of the best writers on bees—yes, the best of them—was blind from early youth; it's as if to show that here, where you would have thought external observation so important, he had found that point and from it was then able by purely mental activity to infer back to all the particulars and reconstruct them in analogy with nature.

There is much to ponder in this compressed paragraph, including Kierkegaard's admiration for "pure mental activity," freed from any empirical bias (or even basis). Nature, for the philosopher, is analogous with God and is far too enveloping and opaque to be fathomed. He becomes lost in the details and can infer the sublime composition only as a whole. Art, however, is specific and legible, by virtue of its being authored by a humble human being, no matter how powerful its aesthetic effect.

The part that tickles me most in Kierkegaard's meditation is his almost defensive insistence on who can be reckoned the best apiologist in the world. (For his money, the pioneering Swiss entomologist, François Huber, author of the 800-page monument *Nouvelles observations sur les abeilles* [1792].) Huber was blind from the age of fifteen and died at the ripe old age of eighty-one, in 1831, in the arms of his beloved daughter. Contrary to Kierkegaard's depiction of Huber as almost channeling his understanding of the bees in a mysterious mystical process, connecting hive with brain with proxy fountain pen, the diligent scientist in fact relied on the eyes and hands of his wife, Marie-Aimée Lullin, and also of his devoted servant, François Burnens. Thanks to this trivial moment of repetition—by a thinker famous for understanding the importance of such—I now amuse myself at the thought of Kierkegaard's almost coming to fisticuffs in the local alehouse, at the suggestion of anyone other than his esteemed Mr. Huber's being considered the supreme expert on bees.

February 24, 2022

Seemingly out of the blue, Russia has decided to invade Ukraine. Apparently, we all need one more reason to lose sleep at night, particularly the poor Ukrainians, who are waking up to a nightmare that seemed, to many, a bad memory from a previous century, rather than a terrifying present reality. One wonders how long Putin has had this latest act of local aggression in mind; perhaps it was put on hold during the pandemic. And now we all must contend with his latest attempt to rebuild at least the core of the Soviet Union for a new geopolitical epoch.

My Ukrainian contacts on social media are obviously in panic mode but have resolved to stave off the aggressor. Some, however, are already seeking asylum, one way or another. I immediately contacted various administrators at my university in the hope of some practical, urgent lines of support for scholars or students in imminent danger. But even the former "university in exile" is hamstrung by bureaucracy.

Yet another way in which we academics prove our prowess in talking the talk, while forgetting how to walk the (much more important) walk.

February 28, 2022

Dear bees—it appears that we COVID-ravaged humans are not the only creatures with access to a new vaccine. Indeed, I just read in the newspaper of record that scientists have developed a "shot" to protect honeybees from American foul brood, "an aggressive bacterium that can spread quickly from hive to hive." Apparently, the vaccine will be delivered via royal jelly, so you don't have to worry about a sore spot on your little buzzing bodies. I pray only that this is more effective than previous "treatments," which included "burning infected colonies and all of the associated equipment, or using antibiotics." Here's hoping you are all spared the scourge of American foul brood. (Which, let's be honest, sounds like possibly the worst kind of foul brood one could be infected with.)

March 1, 2022

Speaking of viruses, I found this orphaned snippet in my notes file, during the height of the first lockdown:

After the last human died, in 2023, the virus moved on to infect other, smaller mammals. Whereas capitalism, now deprived of its main host, slunk off to bide its time, buried deep in some of the last remaining glaciers. There it waited for millions of years until some of the surviving insects evolved and became infected in turn by the profit motive. Soon enough ants were charging each other for small pieces of fungus, and bees were marking up the price of their honey. Worker bees tried to organize unions, but these were busted by voracious agents of the queen. Wasps reinvented advertising, and beetles became some of the most active brand-ambassadors around. Ladybugs figured out how to monetize their feminine charms. The hedges were brimming with tiny, motivated hedge-fund managers. Eventually, seeing the devastation all around them, some aphids lobbied to create a vaccine against capitalism, but the mosquitoes—rabid anti-vaxxers—thwarted such attempts. Eventually, the influenza of influencers killed all the insects as well, so that capitalism was obliged to skulk away once more, this time into the ocean, where—after a million-year hiatus—it first went viral among the mollusks . . . .

March 4, 2022

Yesterday, I saw a gentleman of a certain age, dressed in East Side–tailored clothing, walking his dog in the Park. In a completely normal voice, he said to his companion, "Try to stay on the path if you can"—as if the dog could understand English perfectly. And perhaps, given the lack of condescension in the old man's tone, the dog could indeed.

March 9, 2022

Today I learned about a medical procedure known as BVT, or Bee Venom Therapy. Known also as "apitherapy," this practice dates back thousands of years and utilizes deliberate bee stings to various strategic points on the human body to stimulate the immune system, and allegedly cure a host of allergies, as well as many forms of arthritis. According to *Bee Culture* magazine ("The Magazine of American Beekeeping"): "The use of BVT for rheumatic diseases has been recognized for at least 2,500 years. While the majority of therapeutically applied bee venom is through injection in the form of desensitization shots for people suffering from hyperallergic reactions to honeybee venom (anaphylaxis), anyone with access to a hive can obtain venom from the self-contained, self-sterilizing, self-injecting bee venom applicators living within."

There can, however, be serious side effects. According to the people's online encyclopedia of note: "In March 2018 it was reported that a 55-year-old woman died after receiving 'live bee acupuncture,' suffering a severe anaphylactic episode which the apitherapy practitioner did not respond to by administering adrenaline. While stabilized by ambulance personnel on the way to the hospital, she died a few weeks later from complications resulting in multiple organ failure. Live bee acupuncture therapy is 'unsafe and unadvisable,' according to researchers who studied the case."

March 10, 2022

There's an old-school pizza joint near my apartment: Original Ray's Pizza. (New York City, famously, has several "Original" Ray's Pizza joints, all claiming, of course, to be the first.) The owner of this establishment—of extra-greasy slices and dubious hygiene—double-parks his car in front of the restaurant all the time. I have seen him pull up in the middle of the road, step out of the car, and place an empty pizza box under the windshield wipers, evidently to signal surreptitiously a deal he has with the local cops and traffic inspectors to let them know not to give him a ticket. (A nice instance of folk semiotics and low-key Manhattan corruption.) In the age of smartphones, parking apps, and QR codes, it warms my heart that this humble cardboard box has the same technological power to keep a fine at bay.

March 11, 2022

Do you feel a bit strange today, dear bees? I would not be surprised if you do, given the fact that your entire species has been reclassified. Let me read you today's news, to clarify: "In what is being described as a 'great day' for bees, a California court has ruled that they are fish." The article goes on to reassure us that "biologically speaking, bees are not fish." But in a pragmatic legal sleight of hand—in the name of protecting our beloved pollinators—"the court decided that bumble bees can be classed as an invertebrate, offering them protection under the California Endangered Species Act." True, this applies only to your kind who live on the West Coast, and inside the borders of California. But perhaps New York and other states will follow suit, and we will have hives full of legalized bee-fish, assembling now in schools, and learning about how to collect honey among the salty anemones of the ocean.

Linnaeus and his legacy be damned!

April 12, 2022

More bad local news, I'm afraid, dear bees. A shooting spree on the subway. While such an act is appalling on any occasion, this will unsettle New Yorkers all the more, since our grubby undercommons were only beginning to feel like a place we could venture into again, for this troubled and tentative city. Like something out of a horror movie, the attacker apparently put on a gas mask, then opened a canister that filled the car with smoke, before shooting indiscriminately. (Ten people hit at first count, with the body count unclear.)

Once again, we must wrestle with the gun violence that this country now throws at its terrorized population on a daily basis. And once again, *The Onion* must run its famous headline: "'No Way to Prevent This,' Says Only Nation Where This Regularly Happens."

October 16, 2022

Hello again, dear bees. Another long break between tellings.

I've been, I admit, lazy and remissful over the summer. (I'm not sure that second one is an actual word, but I feel you understand what I mean, and I feel it captures the general ennui that has settled into my specific system.) Indeed, I did manage to escape to France, which was both a tonic for the soul and a constant reminder that things globally are still a long way from returning to normal. I won't detail all the issues we encountered, however, since these would be merely the complaints of the privileged.

Perhaps you don't even recognize me. There has no doubt been quite a few generations of your kind between this update and the last. I suspect, however, that Lamarck was right, and there *is* such a thing as hereditary memories. So even if you don't know me from Adam, you still understand—thanks to a mysterious mnemonic-genetic relay—that a human comes to your comb-home every now and again, to give you the news from their world, when they deem it worth the giving.

Indeed, the wider world is suddenly more familiar with the custom of "telling the bees," now that the queen is dead. (No—don't panic! Not *your* queen, but the venerable monarch of the United Kingdom.) Old Elizabeth certainly had a historic run. More than seventy years on the throne! Four score and ten summers, stuck in some formal occasion or another, and struck onto the coinage of the realm. And as that nation mourned, some news outlets reported that the royal bees had been informed, by none other than the Royal Beekeeper. (A man by the name of John Chapple.) "The mistress is dead," Mr. Chapple told his charges, "but don't you go. Your master will be a good master to you." (The master being the soon-to-be King Charles.) I suspect that the royal bees already knew that Elizabeth had finally slipped out of this world, feeling it deep in their royal jelly.

And so, one of the few remaining pillars of stability and persistence—whether you are a royalist or an anarchist—has now gone. The global disorientation continues. And indeed accelerates. While monkeypox mercifully (fingers crossed) seems to have waxed and waned in a matter of months, COVID continues—and continues to mutate—even as most of the world's population prefers to act as if we are living in post-pandemic times. Humanity, which held its breath for more than two years, has now resumed breathing, albeit with wary and compromised lungs. The general limbo continues. Supply chains have not recovered, and perhaps they never will. The economy (that sinister abstraction) totters and stumbles. Rent and inflation soar around the world, and yet no one has any money. How does that even work? Who is paying so much for so little? And how can they afford it? The war continues in Ukraine, and we try to stop our ears to the fact that the fate of all living things on the planet now depends on the mood and ego of a single Russian narcissist. Then again, we've managed to efficiently commit ecocide without the aid of nuclear weapons. Scientists announced yesterday that 70 percent of the world's wildlife has been wiped out in the past fifty years. (That is to say, in my own lifetime.) How to absorb that kind of information without going mad? After all, it's not just another depressing little factoid, along with all the others. It's not something you can simply scroll past and say under your breath, "What a shame," and then make another cup of tea. And yet, that's precisely what we do.

(Sorry if I'm bumming you out, dear bees. But I'm afraid there's not an abundance of good news these days. Perhaps if I give you the horrible headlines today, however, I can bring you more heartening tidbits in the coming days.)

Floods in Pakistan have killed thousands and displaced millions. Australia's most populated coast is similarly under water. Iran is on fire as a result of protests against the murder of a young woman who refused to wear her *hijab* in public. Fascists now hold power in Italy, the cradle of that particular noxious political weed. Soon we will know if the Amazon forest—the lungs of the world—will continue to be amputated at a record rate, or if a more sane government will prevail. Meanwhile, in the United States, we wait in fear for the second shoe to drop—specifically, a shiny gold loafer. Whether the orange tyrant takes power again or not, the damage has already been done. The Supreme Court has now made it possible to ban a woman's right to access abortion, and the misogynistic states have already started criminalizing attempts to end unwanted pregnancies. (No matter the circumstances.)

Closer to home, here in New York three bodies have been found in my neighborhood in the past two weeks: two men in Central Park lakes and one woman in the Hudson River. These are not suicides, apparently, so the specter of a serial killer has people on edge. The city seems to be recovering, in a superficial sense, as tourists flood back into Manhattan. But businesses still struggle to stay open, residents scrap to pay rent, and parks return to the old days of being open-air crack dens. (I noticed that some enterprising buskers are also selling drugs, to supplement the income from their music.) All this churning turmoil might be even vaguely exciting, if there were some sense of a reboot, or reinvention. But instead, we have the grim realities of the 1970s and '80s, but without the affordable prices, interesting establishments, and vibrant art scene of that grimy golden age. Corporate blight spreads throughout everything, like fluorescent black mold.

Yes, dear bees, it's hard to hold on to hope in such times. The prognosis is not good. (The main reason, perhaps, I stopped coming to deliver you the news for so long.) But we can't simply ignore it. We must, as one wise woman said, "stay with the trouble." We need to acknowledge the crisis, so that we may figure out how to respond to it (whether that be politically, practically, socially, spiritually, or psychologically).

In any case, I hope I have not soured your honey today. As I said, this is a bit of a misery dump, to hopefully clear the way for more constructive thinking, and more inspiring tales, and perhaps even nourishing projects. Even if we're living in the end times, we are still here—for now. And we cannot simply become numb. As a wise man also said, "The system wants us to be sad." So we must fight this *imposed* sadness with our own unruly and creative kind of sadness. A darkness shot through with the gold veins of joys unlived but waiting to be discovered. Even if this discovery is only to be made by others. Perhaps. Eventually.

October 18, 2022

The ongoing media coverage of the death of Her Royal Majesty is reminding me once again of extended sections of Maeterlinck's book, whose protagonist— beyond the whateverbee of typical beehood—must surely be the matriarchal monarch who reigns over the hive in all her formidable fecundity. Certainly, Queen Elizabeth II likely had misgivings about being legally shackled to the odious Prince Philip. But who could say whether she would have preferred the plight of her apian opposite number, inside the combs: the queen bee who is obliged to endure what Maeterlinck describes as "the embarrassing presence in the hive of three or four hundred males, from whose ranks the queen . . . shall select her lover; three or four hundred foolish, clumsy, useless, noisy creatures, who are pretentious, gluttonous, dirty, coarse, totally and scandalously idle, insatiable, and enormous."

Indeed, I entertain myself with images of a counterfactual world in which the recently departed Elizabeth of Windsor lived a resolutely insectoid life. Here, Buckingham Palace is filled with gothic wax hexagons and thousands of buzzing brood-cells. The young queen herself, only a matter of months after the coronation, must, of course—to continue the lineage—venture outside the royal hive, soar up toward the sky with her translucent virgin wings, and engage in what Maeterlinck calls "the nuptial flight." At this time, triggered by some invisible chemical signal, thousands of crooning suitors from the English aristocracy attempt to swarm her royal body, suspended high above the twinkling lights of the capital, tearing at one another's tweed carapaces for the honor of mating midflight with the royal vagina, before being summarily murdered in mid-air and falling back to the ground, deprived not only of life but also of the still-swollen sexual organ which her royal highness—looking more like an exulted maenad than a pensive monarch—will keep inside her enormous ovaries as a bloody and expedient trophy.

As Maeterlinck observes:

Most creatures have a vague belief that a very pre-
carious hazard, a kind of transparent membrane, divides
death from love; and that the profound idea of nature
demands that the giver of life should die at the moment
of giving. Here this idea, whose memory lingers still
over the kisses of man, is realized in its primal simplic-
ity. No sooner has the union been accomplished than
the male's abdomen opens, the organ detaches itself,
dragging with it the mass of the entrails; the wings
relax, and, as though struck by lightning, the emptied
body turns and turns on itself and sinks into the abyss.

Thus (still inside my surreal daydream) when Queen
Elizabeth returned to her palace, inflamed by bloodlust
and several months' supply of semen, she sees the returning aristocrats, their jowls drooping in disappoint-
ment at missing out on the sovereign congress. It is
in disgust at their obsequious defeatism that she—to
quote our expert one more time—"will coldly decree
the simultaneous and general massacre of every male."

October 23, 2022

My students tell me that "the ick" is a thing. (As in,
"This guy gave me the ick.") As far as I can tell, it's a
sub-species of "cringe," which itself seems to be one
of the dominant affects of the age. The ick describes
the sudden repulsion we might feel for someone we
initially found attractive, when they say or do something that suddenly seems like a deal breaker, no matter

how trivial or unconscious. Examples are infinite, but they include "bad bitmojis," "snorting while laughing," "frayed underwear," and "patting the seat nearby to invite you to sit." Of course, this phenomenon has existed long before its current naming procedure as "the ick." *Seinfeld*, for instance, was replete with examples of the main characters' ghosting a recent date because of such minor transgressions as being quiet talkers or having somewhat large hands. One example, provided by a student, has been haunting me, however: "Taking too long before jumping into a waterslide." Apparently the ick extends to anything that puts a crush in the context of being at all fallible, hesitant, vulnerable—human. From this angle, the ick is in fact another word for Roland Barthes's famous *punctum*: a detail that "pricks" or "wounds" the psyche, by virtue of its solicitous intimacy and heart-rending immediacy. In this case, however, the experience is reversed 180 degrees, so that this sudden stab elicits not compassion and empathy but rather disgust and contempt. The ick takes hold the moment someone makes a gesture—or fails to make a gesture—rendering them *too* relatable: unpoised, uncool, pathetic. Pathos can indeed cut either way, encouraging a profound connection or prompting a quick mental quarantine process, so we don't become contaminated by the likely contagious helplessness of the other. Increasingly, we're seeing evidence that the latter is the preferred strategy, when confronted with the discomfort of our own existential exposure, as manifested in another person. Where the punctum expands and connects, the ick contracts and severs. Indeed, perhaps it's time for an updated version of Barthes's famous book on photography: "This particular *ick-punctum* [no longer] arouses great sympathy in me, [nothing resembling] a kind of tenderness."

October 24, 2022

Lions, leopards, giraffes, elephants, buffalo, hippopotami, rhinoceroses, crocodiles, zebras, warthogs. Various species of antelope. I've been watching some live-cam feeds from South Africa lately, and all these exotic animals have wandered into frame, jostling for room at one of the local watering holes. A different tab in my web browser shows grizzly bears in Alaska, trying to catch enough leaping salmon for themselves and their pacing cubs before winter closes in. Truly, we sedentary humans can't get enough of what zookeepers call "charismatic megafauna," even as we're not exactly inconveniencing ourselves trying to keep them from extinction.

What currently strikes me about these creatures is the extent to which they sleep only in hurried (not to say harried) snatches, engulfed in a wider current of extreme vigilance. Some of the apex predators—like lions—sometimes manage to have a decent slumber. They must, however, also be on the lookout for poachers. Zebras, springboks, giraffes, and so on are constantly scanning the horizon for potential danger. (I've read about some animals that can even literally sleep with one eye open, like the crocodile, or allow half their brains to snooze, while the other half remains active, like the dolphin.) Animals—especially as they mature—are rarely afforded time to relax, or to play. Their lifespans are shorter than they could be, in part because of the high levels of stress that constitute the shrill tenor of their daily experience: locked into fight-or-flight mode much of the time. Watching these creatures rest their weary heads on the muddy ground for just a few moments, only to jerk back into alertness once more, makes me exhausted by proxy. It also gives me a renewed gratitude for the relative solace we have constructed for ourselves: walls and doors and roofs and lockable windows to keep the roaming shadows of menace away. Humans are, in other words, the animal that created the conditions for a good night's sleep. (Even if we looked to the sloth, koala—and our own domesticated companions—for inspiration, and even as we're also the creature that fails to take advantage of this privilege, because of over-work, faulty brain-wiring, or the hyperactive activity of internal mental predators.)

As the planet's most perverse animal, we humans are indeed becoming increasingly anxious and prone to insomnia, thanks in great part to the relentless grind, omnipresent insecurity, and general chaos of late capitalism. Even as we invented the luxury of a decent slumber—along with its decadent machinery and soft technologies—of mattress, pillow, curtains, herbal tea, lavender drops, and imaginary flock-based inventory.

Today—given the antagonistic conditions of modern life—we may enlist eye masks, ear plugs, weighted blankets, white noise machines, and other sleep aids, up to and including powerful pharmaceuticals. How comic we must look to the antelope—those savannah-based equivalents of fast food—as we climb up into our fluffy beds, in a world bereft of lions or hyenas or other salivating monsters. And yet we still find the precious and hefty gift of sleep elusive.

Walking through the city streets and parks, I see more and more homeless people returning to the conditions of the great plains. They lie in sodden sleeping bags under sparse trees in the rain; or under cardboard boxes beneath the dripping scaffolding that surrounds half of Manhattan like a tetanus-coated exoskeleton. The unhoused may not fear a midnight encounter with a saber-toothed tiger, but they still need to keep an eye out for curious rats, bored police officers, or even sociopathic frat boys. It is of course a disgrace that so many apartments lie empty, as half-forgotten investments or automated tax dodges for the wealthy, while those who have fallen through the ever-widening gaps in the social safety net must contend with living like orphaned beasts of the desert. Astonishing to think how easily we accustom ourselves to this scandal and even fear or resent those who are themselves the ones exposed to the elements, night after night.

All of which makes me grateful that you, my dear bees, have had the evolutionary good sense to build a hive in which to spend the night. While the reality may not be quite so cozy, I like to imagine a pleasantly soporific hum inside the combs, its hexagonal, honeydripped walls gently vibrating with a calming tone, its hardworking inhabitants droning and dozing off together on soft, waxy beds, inside an abode that doubles as a soft-glowing lantern, still warm from the now-vanished afternoon sun, and perfumed with the bouquet of a thousand flowers; their fragrant traces now mellowing into a golden nectar that nourishes warm collective dreams of royal blessings and pollinated adventures.

October 26, 2022

To say nothing of artificial light.

Humans are the only animal that need to illuminate their way through the darkness through non-natural means. Unlike all the rest of God's creatures—who either happily sleep through the night or who have special senses to navigate as freely as we do during the day—we depend on luminous devices. Before their invention, we stumbled after sundown, our eyes adapted only to the reflected rays of the sun. For millions of years, our ancestors cowered in caves, waiting anxiously for the dawn, when they could roam around in relative safety once more. (A tense and seemingly eternal vigil, captured well in the opening scene of Kubrick's classic film *2001: A Space Odyssey*.) Eventually we learned to control fire to the point where we could carry it in front of us on the end of a stick, like landlocked lantern fish. (To this day, English people call flashlights "torches," which—to the American—recalls visions of angry midnight mobs, in search of slippery witches.)

God's first cosmic act, we are told, was the creation of light. Humans would have to wait a long time, however—whether we are measuring in biblical or Darwinian timescales—before we could flick the light switch and de-darken a room on command, like gods within our own domain. For millennia, fire was our only option for pushing back the night—flickering village bonfires and beacons, and eventually shared and distributed domestically as candles. Finally, at night, we could see the faces of our families indoors, lambent in bright fire-light, whereas before, we had to content ourselves with the occasional pale reflection of the moon. And yet candles didn't grow on trees, so such glowing visions had to be rationed and could be taken for granted by only the rich. Kerosene and advanced metalwork then forged the lamp, which was a more reliable companion to light the path, or hang off the rigging of a boat. (The risk of fire always being weighed against the convenience of extended sight, of keeping the darkness at bay.)

The animals must have noticed this ominously increasing harvest of light, the humans reaping more and more luminescence from the materials around them. Nocturnal creatures now had to contend with shimmering towns, and strange, slow will-o'-the-wisps, meandering around above the ground, carried by travelers and night watchmen alike. Little did the animals know, however, how explosive, invasive, and seemingly everlasting such light would become. (Indeed, the idea of "light pollution" would be absurd to everyone except the most Promethean Victorian scientist.) For half a century or so, before Benjamin Franklin flew keys attached to kites during thunderstorms, we floated like serene jellyfish in the gentle, intimate light of the gas lamp, which even outside in the street created a virtual ceiling against the darkness—a soft buffering—rather than banishing the night sky almost completely. With the harnessing of electricity, however, everything changed almost over-night. Light was now as obedient and omnipresent as water, ready to flow from the tap. The intensity of its brightness could be measured precisely and bought by the radiance as timber was bought by the foot. Remark-able to think that a single lifespan bridges the first luxury European hotels wired for electricity and the towering neon waterfalls of 1980's Shinjuku. Today, people under thirty may not have seen the Milky Way, thanks to the blinding effacement enabled by electric light.

No wonder we are so arrogant. We feel like divine beings, lighting our own way through the cosmos. And yet we also know that beyond this paper-thin shell—on the other side of this magic lantern—lies an infinite expanse of darkness. All the more reason, we feel, to shower the Earth with a halo of lasers and turn the planet into a giant disco ball, glittering in space. Mean-while, the remaining critters squint at us performing our awkward dance moves, now too addicted to the harsh buzz of electricity to enjoy those secret ancient pleasures still lurking in the shadows. (Incidentally, we call any unsavory activity "shady" in reference to those skulking folk who used to make a living beneath the elevated train tracks of New York and Chicago.)

And so to you, dear bees, I apologize once again for my kind, and the distress we have caused to all of Earth's creatures, in this case by coaxing and convincing the sun to shine nonstop, through a billion violent eyes—uprooting territories, disturbing rhythms, banishing entire murky worlds in which others happily made their way before we were even a dim glint in a salamander's eye. We humans presume we are ready for our close-up, for the spotlight. But we have not even begun to reckon with the way that our own "enlightenment" was gained at the exorbitant price of holding the night hostage; of holding her prisoner, along with all her countless followers, in a harsh and blinding cell; a vast and shining fluorescent hell.

October 27, 2022

Which brings us to noise—the form of pollution that I take most personally (since I have very sensitive hearing, which itself is hooked up to an especially delicate nervous system). In keeping with our recent theme, the soundscape of the world must have been incredibly peaceful for the past 3 billion years or so, once the Earth's crust started to cool, like a giant pie, floating in space. Before life evolved in the waters and the jagged cracks, the world would have been sublimely silent, save for the occasional volcano venting into the atmosphere, with not a soul to hear it. (And thunderstorms, of course, which must have arrived like some kind of apocalyptic Wagnerian opera, blowing away as quickly as they came.) Eons later, flora began to tentatively exert its presence, and eventually forests covered the land. In this time, trees fell with no expectation of being heard. Eventually the animals emerged from the primal soup and ultimately grew orchid-like organs—that is to say, ears—to better orient themselves in a dangerous world and hopefully avoid becoming dinner for another. Even so, the aural surround must have been mostly tranquil—just the breeze in the leaves, the lonely call of migrating birds, the white noise of the cicadas, the snort of the mama pig, the snore of the dormouse (tucked up in bed under a leaf), and the occasional snarl of the larger cats of prey.

Before humans showed up on the scene—tipsy on fermented apricots, and with a pile of vinyl records under their arms, ready to party—the loudest sounds must have been those of howling wolves and monkeys; or the great sonorous operas of the sperm whales, suspended in the ocean. (Only a hundred years ago, whales could communicate for hundreds of miles, through the welcoming medium of the sea. But then our navies and industrial fisher-folk turned the ocean into a cacophony of booms, churning hums, and shrieking frequencies, stuffing our hyper-intelligent cousins into claustrophobic sonic boxes, only a few dozen meters wide.)

For this indeed was the fate. The Great Amplification. We started with humble designs: to sing around our campfires, holding our hands around our mouths to call others to the merry music-ing. We whistled and yodeled and yelled, searching for caves and other rock formations to echo our sonic reflections back to us. We invented musical instruments, including the gong, which could reach farther than the human voice and summon others toward it. Then the bell, which was not only the voice of an entire village but also an echo of God. The invention of gunpowder allowed a terrible din, enlisted to keep evil spirits at bay, and keep children gamboling beneath fireworks, wide-eyed and with ears ringing. It must have been about this time—with the fabrication of the cannon and the accompanying rumpus of war—that the animals noticed that the world was becoming louder, no thanks to that one inconsiderate species: that troublemaker, that Johnny-come-lately noisemaker—Man. He hollered, he squealed, and he pressed the elements in such a way that he could now magnify his larynx, so that it boomed like the voice of the Almighty. Hailers, megaphones, microphones, speakers, amplifiers. They all allowed this one animal to preside, profess, propound, and perform at greater and greater volume. Perhaps inspired by our racket, the animals responded with a kind of resentful mimicry. Wolves howled. Elephants trumpeted. Bulls bellowed. Parrots shrieked. Dogs barked. (This latter being "the stupidest sound in nature," according to one eminent philosopher, since it is the sound of a once-wild animal trying to communicate with the enslavers.)

Back at sea, enormous steamships lashed the water with giant propellers, churning the waves into a boiling soup. Vast horns blasted a hellish bellow across the froth, like the sound of an impregnated demon attempting to give birth. On the land, cars sputtered, thrummed, and honked in a more modest, but no less insistent, fashion. Trains chuffed and clanged. Cars grew wings and took flight, trailing a fuel-soaked clamor in their wake. Suddenly, conglomerations of humans were now organized to prioritize the noise-making machines: trams, cars, buses, trucks. Bells, buzzers, and alarms cleared their paint-coated throats and sang out their tortured siren songs at the top of their metallic lungs.

Machines that were once isolated to a small island off the coast of western Europe mated and migrated across the world, trailing a billion clattering, shattering, squealing, and squeaking broods of steel upon steel. When these became motorized, factories evolved into massive, vibrating monoliths, shaking the kitchen walls of those obliged to work, and lose their hearing, in the same sonorous spaces. Cranes, bulldozers, jackhammers, power saws . . . an orgy of unprecedented decibels assaulted the land. Subway cars rattled through tunnels underneath the Earth, like snorting subterranean dragons, in search of their share of Hell's frenzy. Televisions spattered and chattered into empty rooms, or for the company of the seemingly mute, like a gossiping parson who never needed to draw breath. Previously sedate homes filled with cackling, crackling radios, whooshing hairdryers, chugging and humming appliances, and shrill, trilling alarm clocks.

Humans, it seems, craved the cocoon of noise and claimed to be unnerved in the presence of its absence. They sought out concerts, theatrics, nightclubs, and hubbubs of all kinds. Having left the womb of a quiet Earth, they were now addicted to having their eardrums pummeled and their nerve endings jangled. They invented headphones and portable music players so that they may walk in the park, their very brains being pummeled with the sound of techno or heavy metal. (That is to say, the sound of the factory, adapted to a rhythm pleasing enough to disguise its laborious provenance.)

As one of the millions of human moths drawn to the flame of Manhattan, I can attest to the ceaseless noise of this, in many ways the opposite of the anticipated "smart" cities that may eventually learn to put a lid on it. New York is still very much a dumb city. And a deafening one at that. Though worse than the infamous ear-splitters—like ambulances, police cars, and fire engines—are the more constant and tiresome noises, the ones that wear away the spirit like steel wool. Leaf blowers, car horns, car alarms, smoke detectors, garbage trucks, ice cream trucks, power tools, generators, renovators, HVAC systems, industrial fans, air conditioners, brick polishers, sewage suckers, hammering, yammering, and all those unidentifiable hummers, beepers, whooshers, thumpers, and clunkers that punctuate the urban symphony with such graceless insistence. For me, personally, the helicopters are the worst—the ones hired by tourists to jaunt up the Hudson River and over Central Park. These are nothing less than flying jackhammers: sloppy jalopies in the sky, revving their engines and squatting over the miraculous gift of the Ramble like a flatulating squadron of airborne Harley-Davidsons, irritating thousands of people on the ground—and no doubt thousands more animals—for the pleasure of three or four moneyed visitors.

I could, of course, escape to the relatively rural arms of upstate, as so many have done recently, in a bid to find loams of silence, just waiting for me. But I also know there's always a barking dog, or a buzzing chainsaw, or a droning lawn mower, or a thumping sump pump to banish peace of mind. I also realize that despite the din of the city, I need to cling here for a while longer. (Not that I have a choice, economically speaking.) This dissonance is terrible for the body, which must spend vast reserves of energy to screen it out, to metabolize it, so that we are not driven mad. (Madness itself causing the nightly sound of grinding teeth.) But the spirit continues, amongst the sonic assault.

I wonder if this little fable strikes a chord with you, dear bees. I wonder if your hive is experienced as noisesome, with all that buzzing. Or is that buzzing all you know, and thus the functional equivalent of blissful silence? (I do suspect that if the buzzing were suddenly to stop, it would send the hive into a terrible panic.) Which is all to say, the only way to really escape noise is to cease existing. (For even the deaf "hear" through their skin, and through vibration.) So, noise is the price of being. A price now pegged to runaway inflation.

And so, for the third day in a row, let me apologize for my species and, in this case, for the commotion that we have visited on all those who cohabited at civilized volumes for so long, with no need for noise inspectors and white noise machines (an ironic device, being a machine that makes a sound to mask that of other machines) and noise-canceling headphones. My only consolation is that we are also the animal that has invented deafening weapons of such magnitude that we may return, sooner than we think—and after a great bang precluding all whimpers—to a planet of quietude. One hopes, however, that the world (and by "world," I mean all the other critters, fungi, and other intelligent elements) learns to unplug these deluded, wannabe rock stars (the humans) without taking all the other living things—including you, my dear bees—with it.

October 28, 2022

What comes first, pain or the one who feels it?

The answer seems obvious. The latter. For we first need a being to exist, in order to have pain visited upon it. But what if we have it the wrong way around? What if pain already exists in the cosmos, like mushrooms under the soil, and awaits occasion for its blind and shining moment? What if the Big Bang contained Everything—including pain—in its impossible density, now splashed across the entire universe? Milky, mercurial latencies patiently awaiting their potential to become manifest. Perhaps life was summoned by the cosmos for its sensitivity alone. (Suggesting we somehow miss the point when we rate sapience higher than sentience.) Perhaps everything with the capacity to feel the effects of the universe upon its sensorium—whether it be the breeze, the temperature, the light; or, indeed, pain, pleasure, relief, release—is a sensor pre-designed to register these events precisely. Like a rainbow—which exists nestled in the invisible folds of the light spectrum, only to be released and revealed by certain atmospheric conditions—pain, pleasure, and all the other affects, lie curled up and in wait.

Yes, the universe is cold and dark and empty. But there are billions of possibilities for exceptions to this rule, albeit scattered millions of light-years apart. Blossomings of something other than unfeeling. Blooms of exemptions from the mute general law of impressionlessness. In this case, the human nervous system is an advanced organic technology for detecting stimuli and transducing it into intensities of differing degrees and qualities. What we call "emotions" would thus be the auto-registration of phenomena that already existed but were waiting for us—or something somewhat like us—in order to emerge; in order to show up for their fifteen minutes of cosmic fame, as it were.

From this disorienting perspective, pain indeed precedes the afflicted. And the same could be said for joy. All of Earth's creatures would thus be sensors created by the universe so that it may have a deeper and more nuanced sense of itself, and its own effects. An almost infinite—and infinitely distributed—cosmic phenomenology.

I hope this stray thought gives you, my dear bees, something to contemplate, as you go about your daily business today. May it be completely pain-free—even as you carry suspended agonies in the sting of your tail.

October 29, 2022

I've been reading a new book that I think you might appreciate.

(I know I've been speaking to, and about, "you" all this time, as a kind of collective individual, as if all you different bees in the hive might be interested in the same things, and even as I know "you" can have very different characters, roles, behaviors, interests, and so on. So, forgive me if I continue to address you in a kind of singular-plural, for the purposes of this experiment in unidirectional interspecies communication.)

In any case—and as I was saying—I've been reading a new book. It's a kind of philosophy book, but one that is trying to approach the question of ontology—or Being itself—from a disorienting new angle. (The title is rather lovely: *Bizarre-Privileged Items in the Universe*, itself a phrase borrowed from the French polymath Roger Caillois, who was also very interested in insects.) The book's author, Paul North, argues that we have been thinking about the world, and its almost infinite inhabitants (both animate and not), from the wrong angle. For while we usually presume that the world is populated with "things"—that we can then describe, compare, contextualize, and so on—we have been missing the forest for the trees. And the forest, in this case, is the primacy of the various traits of any given thing. (Say, color, shape, texture, morphology, or what-have-you.) In other words, we should be starting with *likeness itself*. We should be following the ways that resemblance—as "the basic element in this cosmos"—*co-constructs* the entities that then appear to share certain traits. The upshot, quite simply (for North, at least), is that there is no such thing as a thing! Instead, there are only temporary refuges, or way stations, where likenesses gather to express themselves in precisely the kind of cosmic mimesis that so fascinated Caillois.

I can't pretend to understand the whole book, since its erudition is deeper than mine and delves into some extremely technical and esoteric texts from the long history of philosophy. But if I have the basic argument right, we should not be confidently asserting—in the manner of common sense—that there are bees, and eagles, and airplanes, some of which share certain characteristics (for instance, wings). Rather, there are certain likenesses, such as shapes, in the repertoire of the universe (for instance, wing-like geometries) that attach themselves to different occasions. As the author puts it, "Butterfly and leaf are alike; they enliken one another. From this position, other potential likenesses suggest themselves. Take a bumblebee, a school bus, and a journalist. The are alike in yellowness. Alike in carrying and transmitting important things. Alike in punishing you when you misbehave." In short, likeness comes before beings (or being). Or before even form itself. (The point in the argument where Plato would no doubt roll up his sleeves in preparation for a vigorous swing.)

The purpose of such attachments and occasions? The drive behind all this promiscuous "enlikening"? Well, this is the part I haven't got to yet. (I'm only halfway through the book.) But the premise is startling. And I admit that I do love to have my fundamental assumptions about "the order of things" shaken, even if it's impossible to avoid lapsing into the more common ways of navigating, and understanding, the world, once I close the book and start making lunch.

Something to consider at least, dear bees, as you go in search of flowers that sometimes rather shamelessly pretend to be something they are not, for the bizarre privilege of being ravished by your eminently seducible kind.

October 31, 2022

Today is Halloween. The day that Americans celebrate the horrors of Type 2 diabetes.

The big political news of the day is that Lula won the Brazilian election, hopefully toppling that despicable tyrant Bolsonaro. (I say *hopefully* because there's no guarantee the latter will hand over the steering wheel of the country without a fight.) The victory margin, as always, around the world, was less than 1 percent. And it will never fail to perplex that essentially half the population of any given state will not only vote against their own interests but also happily pull the lever for a sadistic kind of collective suicidal death wish. (Bolsonaro is intent on razing the Amazon for short-term profit, dooming the future of the planet and all its inhabitants in the process. Among a litany of other crimes.) One indeed wonders why a single human being—whether it be Bolsonaro, Putin, Trump, Johnson, Xi Jinping, Modi, Erdogan, Orbán, Netanyahu, Meloni, or whoever will rise from the slime next—can have so much power, so much leverage with the world's vital organs, holding us all hostage because they weren't given enough cake as a child (or, more likely, given too much). Or because they were rejected by an attractive bookish type in the college cafeteria, as a youth.

Catastrophism, I admit, has been a hobby of mine since I was a young lad, long before I heard the term for this condition, for this would-be prophylactic form of magical thinking. ("If I can think of the worst thing that can happen, then the gods will prevent it from happening, since I was not so arrogant as to expect things to go in my favor.") The problem with this rather common affliction—this preemptive coping strategy—is that the catastrophist still lives through the trauma of the event, whether it happens or not. (Or at least a private simulation of the event, which still wreaks havoc on the body, which prepares itself chemically and mechanically for the situation to imminently arise. The fight or flight, once again, becomes locked into the "on" position.) As Caesar, or some other Great Steward of History, once putatively said, "The brave man dies but once; the coward, a thousand times."

Which is why, dear bees, I'm trying to practice "the power of now": a modernized form of Zen, popularized by the New Age guru Eckhart Tolle. (As already mentioned, in the form of his local proselytizer.) This approach or attitude attempts to banish anxieties about the future by simply not thinking about it, by dwelling in the present, rather than in the future imperfect. The mantra is simple: There is a power and presence in the moment you are experiencing right now, and the more you attend to it, the more you realize that tomorrow literally never comes. You *always* live in the now. So why spend so much time chewing your nails about a future that will not arrive? You are Zeno's arrow. We are *all* Zeno's arrow—heading toward something we won't technically reach. (Simply because "now" will always absorb and supersede "soon.") So, if you recalibrate your entire mental, emotional, and physiological system toward honoring and inhabiting *right now*, then you are much better prepared for that mysterious transition from "that which is not yet" to "that which is undeniably here." (And in this sense, it's a strategy attached to the flip side of the classic refrain "This too shall pass.")

Of course, being of both academic and pedantic mind, I worry that this has an inbuilt flaw. Specifically, what if "the now" sucks ass? What if, for instance, you are getting a filling at the dentist's office, or your beloved is breaking up with you, or you are in prison, or strapped inside a plane in heavy turbulence, or the nuclear sirens are starting to sound across the city? What if, as we learned with COVID, "now" can lose its integrity and ooze into a kind of temporal jelly with no shape, purpose, direction, or momentum? How can "being in the now" help when we are precisely suffering from life's seemingly eternal (and increasingly infernal) now-ness?

No doubt, Mr. Tolle has already been confronted with these questions. And I'm sure he has a ready answer. (An answer I would already know, if I had the patience to read the whole book on which his philosophy is based, rather than the more accessible primer that sells much better on Amazon—the one I, in fact, read.) But overall, I think it's a good mantra to have in one's pocket. After all, why torture yourself with hypotheticals that may or may not come true, when the present is challenging enough? (Although, as Tolle insists, the present moment is rarely as challenging, in practical or visceral terms, as we fear.) Anxiety is the fear of nothing in particular, or of something you can't name or identify. So perhaps Tolle is right: The key to keeping anxiety at bay is to stop looking for monsters to pin our fears on, and stare Time right in the eye. "I see you, Time! Don't devour me just yet! For I am appreciating the gift you gave me. I will not squander it by reliving embarrassing moments from the past, or painful scenarios in the future."

November 2, 2022

And speaking of all things Zen—or Zen-adjacent—I've been good about practicing *qigong* every day, a habit that truly saved me during the lockdowns, helping metabolize much of the stress of that period and keeping my body from freezing into a doom-scrolling beetle-husk. (Locking up from being locked down, as it were.) I became a virtual disciple of Robert Peng, a *qigong* master who eventually left China for upstate New York. During the first year of COVID, Master Peng managed to "pivot" to video and taught his accessible classes to the masses for a reasonable fee. He has the most reassuring, avuncular manner, with a charming tic of saying most things twice, as if every observation comes with an identical twin, for emphasis.

I was beguiled by Master Peng's lo-fi tech setup and grassroots approach to the whole endeavor. He seemed to film all sessions in a modest spare room, with a green screen behind him, where open-access footage of foaming waves would splash behind him, or public domain wallpaper of soothing forests. Despite the digital delivery, the connection felt very "real." (In terms of being heartfelt, sincere, and—perhaps most important of all—affordable.) Master Peng would walk us anonymous acolytes painstakingly through the twelve movements of the Yi Jin Jing—a foundational cycle—activating many of the major vital organs and demonstrating how to perform profound biological, psychological, emotional, and spiritual transformations. These happen (or almost happen) through a kind of graceful alchemy; examples include making anger "melt" into enthusiasm, or grief "dissolve" into kindness. For me, punctuating each day with *qigong* felt like a lifeline back to both the wider world that had been so suddenly snatched away and my former self, who lived so naïvely in a timeline that did not overtly resemble an excruciatingly dull horror movie. It cleared a preemptive healing space during the panicked claustrophobia of 2020 and into 2021. I could visualize myself as the center of the universe—as the "axis *mundi domini(c)*"—with my feet stretching into the Earth, like ancient tree roots ("a tree grows old by its roots," insists an old Chinese saying), and the top of my head sprouting branches, scraping the ceiling of the sky. Thanks to the generosity of Master Peng, I can now push mountains off to the horizon with outstretched palms; I can lift burdens heavier than the one troubling Atlas himself and stretch my spine so that it becomes an articulated ivory column, connecting heaven and Earth. All for the low, low price of a couple of cups of coffee a week.

Sadly, Master Peng's efforts were noticed by some predatory New Age entrepreneurs, and he essentially became the target of an aggressive takeover bid; now accessed only for ten times the former price, and only making appearances at glossy WASPish "summits" and other power-Boomer junkets and networking retreats. I suspect he's even parachuted into Martha's Vineyard to give personal lessons in cosmic stolidity to Hillary Clinton and her disenchanted minions. In any case, I was lucky to have a glimpse into Master Peng's world, and his teachings, before it became a brand; when it was still a type of instinctive "good works," with less-polished production values than your average home-baked YouTube video.

As a college professor, I've been teaching a couple of courses that include some moments, or elements, of meditation, since it is, after all, the first technology developed by humans to pay attention to attention itself. As such, it is both timeless and cutting-edge. The stampede toward "wellness" in the general culture is, of course, more a symptom of the privatization and individualization of care than anything else. The more the government and social services—and even family members—abandon us, the more we are left to our own devices to cope with life without a safety net. Accordingly, corporations love to forfeit substantial support programs—like quality health-care coverage, or comprehensive child care—suggesting instead easy tips and affordable apps for "centering ourselves" and practicing "self-care." (Subtext: "Because no one else is going to care for you.") The whole charade has been well described as "McMindfulness."

Much older techniques—like yoga, *tai chi*, and *qi-gong*—can certainly be cynically appropriated and deployed to keep the peasants from revolting, or simply burning out. But they also endure on their own terms because they provide a time-tested way of enhancing the negentropic trick of being a sentient being. They help us realize our potential, against a formidable attention economy (and media ecology) that does everything in its power to keep us from using our brains in a way that prioritizes personhood over profit. Even before the invention of the Internet, as the over-thinking animal, we tended to push our intelligence full circle until we began to resemble twitching squirrels. At the same time, we succumb to the hypnotism of the screen, falling into a kind of binge-bulimia coma. In between these extremes, *qigong* and its ilk can help us take ownership of our own minds and bodies and choreograph a more harmonious relationship between the two (via the spirit). They are a *discipline* in the best sense: teaching and empowering us to have more agency over—and more perspective on—our thoughts, feelings, motivations, behaviors, and so on. (In this sense, we might include psychoanalysis as a form of meditation, or even an especially sedate type of martial art.)

It was my sister, a Buddhist monk, who encouraged me to take up *qigong* in the first place, and for that I'll be eternally grateful. Her own master is called simply Zhao—another almost mythical being, and "*dharma* brother" of Master Peng. For years Zhao hosted small, in-person gatherings in remote places, where his students would learn to selectively open themselves up to the cosmos before closing the points and portals again so that the fresh *qi* would not escape too quickly. It was, by all accounts, a rigorous and rewarding practice, one worth flying across from the other side of the world to attend. As with Master Peng, in the good old days, such classes cost just enough money to cover costs. By a cosmic miracle, Zhao moved to the same small town on the Australian south coast to which my mother has also, by sheer coincidence, retired. As a consequence of the latter, this is also where my sister found herself trapped during Australia's extreme lockdown during COVID, the virus striking during an ill-timed "brief visit." (My sister subsequently petitioned the government to return to her temple in Japan but was consistently denied permission.) The silver lining of this—admittedly rather idyllic—exile was her sudden, unexpected proximity to her *qigong* master. Much to my sister's chagrin, however (if a Buddhist monk is allowed such an emotion), Zhao decided that he too was ready to retire. And so, instead of leading transportive and transformative *qigong* workshops, one-on-one, Zhao would spend almost all day on the dimpled, limpet-covered rocks that extend out into the Pacific Ocean, fishing for his dinner. As my sister conveyed over the phone, in exasperated tones, "It's like living next to the Buddha, but rather than being able to discuss the ways and wisdom of the world with him, he just wants to go bowling."

Of course, being a true *qigong* master, Zhao is not "simply" fishing. Where Dave, Mitchell, or Tanya might spend all day watching the sinker bobbing on the surface of the water simply to avoid their family so they can drink beer in peace, Zhao is surely absorbing all the *qi* from the ocean—morning after morning, evening after evening, in marathon meditation sessions. Like a lightning rod for this specific terroir of the universal energy, Zhao is truly—to quote Master Peng—"a sponge, absorbing an ocean of *qi*." There he stands, fishing rod in hand, the fine line linking him with the waves: a literalized metaphor between what is and what has been. If approached with the right spirit, and in the right frame of mind, fishing can indeed be a supreme form of contemplation, especially when you know how to let all that elemental energy surge into every cell of your body, every particle of your soul. As such, Zhao may frustrate those students he has seemingly abandoned. And he may appear a humble and unremarkable figure to those stray figures walking along the beach, barely registering one more fisherman at the shore. Little do they know, however, that he is in fact a human battery, charging and recharging himself, from dawn to dusk, with one of the most active "outlets" on the planet. As such, Zhao may outlive us all. And one day he may even return to his lessons, restored, and with renewed purpose. Ready to help ground the grandchildren of our grandchildren of our grandchildren.

November 3, 2022

As noted from the outset, there is a presumption in this process, isn't there, dear bees?

In coming to see you and telling you what's happening, what's on my mind. There is indeed an arrogance nestled in the heart of the act of storytelling, since it assumes an interested audience, at least *in potentia*.

Certainly, the last thing the world needs right now is more stories. Everywhere you turn are books and blogs and posts and substacks and memoirs and autofiction. Every other person you meet is either a writer or a would-be writer. A new slew of books appears on Amazon every millisecond. More titles than there are people on the planet. Who is supposed to read this vast and pulsating—and mostly neglected—archive? Who has the time? Or even the inclination? Now that Google has stopped automatically scanning all new books, not even bots bother to ingest what we are offering up for mental consumption.

Which makes me wonder, have I been going about this all wrong, contributing to the problem with my own mini-parade of titles? Perhaps it's long past time I give up adding to the world's choking landfill and start to go in the other direction. Now that I'm heading down the other slope of my life, toward the abyss that awaits, I should not be cracking my knuckles yet again before starting upon a detailed memoir, but rather *undoing* all the work that I've already foolishly sent out into the world. Maybe I should be *effacing my traces*, like a creature in the snow who knows that it's best to slip through the world as invisibly as possible.

How to begin such a project? (Or anti-project.) How would publishers react, for instance, if I contacted them, asking them to cancel and pulp my own work? Would they oblige? If so, would they charge me for the disposal fee? And if not, would I then have to sneak into warehouses and apply TNT to my own books? Then again, most titles are print-on-demand these days, so I would likely have to request that they expunge the listing from all databases. Failing that, I might have to pay a Russian hacker to do the deed for me. Then I would have to task my student research assistant with stealing copies from every library in the tri-state area. (As a prelude to using TaskRabbit to deploy an army of such strategic tome-swipers around the world.) In contrast, they say it's relatively easy to delete one's social media profiles, though one can never trust that a copy no longer lurks on the company's servers. Europe has a "right to be forgotten" law, in which citizens can apply to have their names wiped from the Internet. I wonder, however, if that law applies to holders of British passports, like me, who are no longer protected by EU laws. In any case, it would be a challenge to un-publish myself from the Internet, given how sticky the World Wide Web is.

A strangely appealing prospect, though, to spend the first half of one's life asserting some kind of public presence, to feed the will-to-recognition that we all drag around with us like a limp, and then, for the second half (gods willing), diligently repenting for such misguided narcissism and erasing all footprints, fingerprints, and traces of oneself until there's nothing left but one's own body again, as when we were but mere babies, a few hours old, with blank smiles and even blanker CVs.

November 5, 2022

How strange—to be sitting here in the Park, in my usual spot, wearing a t-shirt. The sun on my bare arms starts to sting in less than a minute; it's practically unheard of for the weather to be this warm, this late in the year. The Park is covered in security fences and temporary structures for the annual marathon. And this year, the runners—arriving from all four corners of the globe—will be competing with summer-like temperatures, as much as with one another. (Apparently belated hurricanes in the Atlantic are swirling warm air up from the South, which always feels disorienting, unnatural.) It must be confusing for you too, my buzzing bee friends.

This weather is certainly a boon for those industries relying on the first real influx of tourists since the advent of COVID, especially the restaurants, most of which have improvised patios, terraces, or shacks floating outside their establishments, like hastily built life rafts. (Which is essentially what they are.) Locals are ambivalent about these omnipresent, all-year-round, nondenominational *sukkahs*. On the one hand, they are a tangible sign of what it took for most of these places to stay in business. (Even as so many establishments weren't so lucky over the past few years.) They provide an ad hoc facsimile of a European city, where citizens take dining *al fresco* for granted. (Certainly, prior to 2020, diners in New York were almost always sealed up inside buildings and behind glass, since the "natural" flow of the street was too precious to be compromised for mere human commensalism). On the other hand, these shantytown structures tend to be ugly, soggy, and smelly, as well as a festival tent for increasingly brazen rats. These idiosyncratic gazebos have annexed much of the street and provide blind spots for the cavalry of deliverymen (for they are almost always men), zipping from place to place on *Mad Max*-style e-bikes.

Whether one is for or against this sudden transformation of the public sphere, it is certainly a radical mutation of our streets, giving the impression of a welcoming, open commons, even if these splintered and graffiti-covered pagodas are in actuality a necklace of privatized bubbles. (It is, of course, necessary to be a paying customer in order to sit in one of these huts.) Still, on balance, I'm personally happy to cohabit with these new outposts—at least for the time being—since they symbolize the resilience that this city is so rightly proud of, including the impulse to extend and accommodate, rather than withdraw and close off. The debate rages on, however: According to one view, these patios are a crystallization of the generous and indomitable human spirit to convene and converse; the other side insisting that they instead represent the belligerent human spirit to indulge in bottomless mimosa brunches, no matter the ultimate cost to others.

As such, the new human hutches are monuments to the ongoing weirdness of the times, in which things slowly move toward a return to normal while simultaneously aggravating a strange limbo. The city, the people, the world—we're starting to get over, or past, the pandemic and live with the virus less as a monster and more as an unpredictable, and potentially dangerous, neighbor. But there is clearly no simple "return" to the before times. When I now go to my campus, most of the doors are closed. Corridors are mostly empty, except for a lone student, wandering masked between classroom and bathroom. Staff work from home, occupied with invisible meetings assembled in the ether of Zoom, and faculty still prefer to scuttle in to teach, deliver their classes, and then scooch back home again. A third of the people I see on campus (which doubles as Greenwich Village) are wearing masks, another third are not, with the remaining wearing masks beneath their noses or chins. So many of the places that were vital in making the place feel like a neighborhood— like a community—have gone out of business, turned into pasted-up storefronts, with vast staging-stations behind the façade: logistical distribution centers for DoorDash, Seamless, Grubhub, ThingBring, and other opportunistic businesses. (OK, I admit I made up that last one.) Indeed, it's remarkable how "disruptive" delivery apps have changed the city itself: rushing in with extra energy during lockdown; uprooting established communal hubs and hollowing them out into anonymous, disguised warehouses. Now, there's nowhere to sit and have coffee; nowhere to browse for a book to put on next year's syllabus; nowhere to meet a colleague for an affordable lunch. (Nothing, that is, beyond the usual fast-food franchises, which I suspect will survive the Apocalypse itself.)

In these diminished conditions, I wonder how long the tourists will keep coming, if there's nothing to see but banks, pharmacies, vape stores, new Hoovervilles, and block upon block of opaque store windows, behind which are hidden anthills of home-convenience commerce.

November 12, 2022

A notable moment in my scanning of the news this morning: Scientists believe they have evidence that bees like to play. Bumblebees especially, apparently, when given the chance, "like to fool around with toys." These same researchers are now asking themselves if bees have feelings; whether they enjoy a rich inner life. The most amazing thing about these "findings," from my perspective, is not that bees will make detours to play with little wooden balls, or enjoy wiggle-waggling through an obstacle course, but that we humans are so amazed that this is the case. Again and again, despite the endless evidence in front of our eyes, we patronizingly ask ourselves if animals are in fact more than simple organic robots, prisoners of instinct. We are so arrogant about our own reputed intelligence, and emotional range, that we will acknowledge some of the same only for the animals that most resemble us; and even then, we do so begrudgingly and with extensive reservations and qualifications. The ghost of Descartes—who believed animals to be essentially animate clocks—still haunts us.

November 18, 2022

Biting wind today. As if it just remembered its contractual obligation to blow off all the leaves by Thanksgiving.

December 1, 2022

Today I bring the sad news of the death of Christine McVie. While hers may not be a household name, most people would know at least one of her songs, as she was one of the main members of the supergroup Fleetwood Mac. Many people my age and skin color grew up in a house with a copy of the album *Rumours*, which was so ubiquitous in the 1970s and '80s that you might be forgiven for thinking that the government simply shipped this sonic staple to every white-collar family in the Western world. McVie's track "Songbird" was the standout ballad on that record. And while it has since become a wedding day cliché, there's no denying that it's still one of the most beautiful melodies ever composed. (Indeed, McVie herself has explained that the whole song came to her in the middle of the night, as a sudden and complete aural vision, and she was terrified that she might forget the haunting tune before committing it to tape.)

While I confess that, as a youngster, I was much more interested in the more obvious glamorpuss of the band, Stevie Nicks, I can't help but feel a twinge with each and every passing of these sonic architects of my own affective universe. (A personal universe that is itself simultaneously collective and shared, since it is an artifact of "the structure of feeling" of the wider mediascape: These songs form the connective tissues of our emotions, experience, memories, hopes, and so on.) Indeed, enough time has now passed from this heroic age—when the mighty dinosaur bands stalked the Earth—that I wonder if any of the celebrated bands of the time could even reassemble their original line-up if they wanted to. Given the ongoing crash of the demographic wave, these legends of the Baby Boomer generation will be dropping off with increasing regularity in the coming years, leaving less and less personal connection with the gestalt soundtrack that formed the communal score of our lives.

We've already lost David Bowie, George Harrison, Prince, and many others, to the voracious appetite of Chronos. It's only a matter of time—itself seemingly accelerating, like a river rushing faster toward the waterfall ahead—before these dearly departed are joined by Neil Young, Joni Mitchell, Bob Dylan, and other names haloed by legend: our postmodern bards, our familiar faces carved into Mount Olympus. And while it seems absurd to grieve people whom you have never met, the magic of modern technology means that favorite musicians have meant more to us—been more "present" and more consistent in our lives—than many friends and family members. They have sung us to sleep, wrung tears from our stony hearts, filled our soul-sails with war cries, inspired us to great feats of courage and unexpected flights of creativity. We have an intimate knowledge of the grain of their voices, which are perhaps a more revealing aperture to the soul than the eyes, even if we know very little of their minds and moods.

Without wishing to sound like a Boomer myself, will we, as a culture, mourn the loss of Taylor Swift or Kendrick Lamar to the extent that we saw with John Lennon, Elvis Presley, and Michael Jackson (and will, soon enough, with Paul McCartney, Keith Richards, or Kate Bush). The stadium-filling bands of the 1970s and '80s were the product of a new media ecosystem that relied heavily on a narrow band of broadcast channels, like radio and TV. Today, with the kaleidoscope of publicity vectors, and the seemingly infinite choice of streaming options, we no longer have the same canon of artists forming a shared soundscape. Even as today's sales numbers may dwarf those of the twentieth century, the initial cohesion of that culture could not hold. Any young music lover we may see on the street today is likely to know only K-pop, or alt.folk, or some Christian country-rock star who commands millions of followers but whom the rest of the country could not pick out of a police lineup.

In short, there was something unique about the postwar generation of musicians, even as there were, of course, all sorts of baroque taxonomies and hidden subcultures beneath the usual suspects of national FM radio play. The point is, no matter how much you may have eschewed The Rolling Stones at the time and sworn fealty to the Velvet Underground, Moondog, Serge Gainsbourg, or Sun Ra, you still lived and breathed within the inescapable medium of America's Top 40, or Britain's *Top of the Pops* (or your local equivalent). Even the edgiest hipsters had their ears partly shaped by the most popular songs of the day. (And thus, their sensibility, as well.) This is why someone like Pete Townsend, of The Who, admitted to getting somewhat weak-kneed when meeting Björn from ABBA.

As the Boomer generation starts to thin out, like the hair on Ringo's head, we're losing precious individual instances of an unprecedented artistic explosion, even if their last moment of inspiration was before the turn of the millennium. After all, almost everything we hear today has its roots in the rock 'n' roll of the 1960s, '70s, and '80s. (Even as these were themselves riffing on early musical genres.) And just as it means something more significant than we can really express, when the last soldier to fight in World War II breathes his last, we'll also lose something almost sacred, when the last great musician of this epoch sings his or her final note.

December 2, 2022

I was thinking, dear bees, about the idea I raised yesterday, about "the last Boomer." Certainly, a figure not as easy to identify as the last surviving soldier (of any given conflict). Where is the cutoff point? Who is still a late Boomer, rather than an early member of Generation X?

(Neither to be confused with a late bloomer.) Is Heart, for instance, a dinosaur band? Dire Straits? Is Bruce Springsteen to be spoken of in the same breath as Mick Jagger? Can we squeeze Rush into the same booming tent as Pink Floyd? These are largely debates for music wonks in pubs, who enjoy sparring around the narcissism of minor differences. But it also raises interesting issues about the simultaneous significance, and cultural conceit, of naming specific generations and, by that gesture, attempting to create some kind of cohesion within them.

The Boomers have also been on my mind lately, as I try to work my way slowly through Peter Jackson's nine-hour archival endurance test, *Get Back*: a fly-on-the-wall, seemingly endless glimpse of the Beatles' creative process, during the recording of their last album, *Let It Be*. A few friends and family members have noted how intimate the whole thing felt and gave them a new appreciation of the genius of what many consider to be the world's most important band. And while I agree that they wrote dozens of brilliant songs, I admit I skipped ahead to part three, so I could get to the climax. (The famous rooftop concert, interrupted by London's overzealous police force.)

Perhaps, rather than the last Boomer, we should simply create a monument to The Unknown Boomer: an abstract postwar person-without-qualities who can symbolize the entire generation. Here, the younger folk can pay their respects—especially for the legacy of their music—while also venting their anger. Indeed, we shall have to budget for twenty-four-hour security for such a monument, given the resentment toward this blessed demographic, who enjoyed more wealth than any preceding group and who—for reasons known only to them—collectively decided to pull up the ladders and squander the wealth, rather than pass it down to their kids and grandkids. (With many individual exceptions, of course.) Sadly, this isn't just a hackneyed narrative, perpetuated by the younger folk, but a stark economic fact, supported by endless graphs, data, charts, and so on. As such, The Unknown Boomer will represent the possibilities realized by our evolving attempt at civilization—during a fleeting window of time, and in very specific pockets of the world. The monument will silently say, "Sorry for taking all the cake and trashing the joint. But here's our musical back catalog as compensation." It will also be a quiet place to reflect on the deeper reason why such an adventurous, creative, liberating demographic force imploded and retreated into the gated community of the self. In the meantime, we have the graves of specific celebrated Boomers to function in a similar way.

Which reminds me of an amusing exchange that happened to me a few years ago, while walking through Central Park. A European couple, perhaps Italian, gestured for me to stop, so they could pose a question.

"Where is the dead beetle?" they asked, in very thick accents.

At first, dear bees, I thought they were inquiring about one of your fellow insects, and my instinct was to point all around.

"Dead beetle?" I replied. "Probably lots of them. On the ground. If you look close enough."

After a lot of pidgin English and general confusion, I realized they were trying to find Strawberry Fields. That is to say, the memorial to that beloved "dead Beatle," John Lennon.

December 3, 2022

I was contacted today by a high-concept "ideation" agency in Germany, inviting me to give a two-hour "expert talk" on the difference between *joy* and *fun*. The (*very* brief) "brief" is intriguing enough to get my mental wheels spinning. What are indeed some key differences between joy and fun? Without reflecting on it too deeply, I figure that joy is deeper, older, and more closely connected to spiritual experience, like grace, ecstasy, and so on. It feels both more personal and numinous at the same time. Fun, in contrast, feels like a modern phenomenon, something engineered, usually in a group. Fun seems more active than joy. It also comes with more of a promise than a guarantee. (Hence the famous phrase "Are we having fun yet?")

Indeed, Adorno—that notorious curmudgeon—was no fan of fun, writing, in *Dialectic of Enlightenment*, "Fun is a steel bath." What this means exactly is not crystal clear, but it certainly doesn't sound—well—fun. He goes on to note that "the pleasure industry" never fails to prescribe a stiff dose of fun, since "It makes laughter the instrument of the fraud practiced on happiness." I doubt, however, that a group of brand managers want to hear what their sour compatriot had to say about fun. Better, perhaps, that I try to at least sneak in a quote by his contemporary reincarnation, Grumpy Cat, who confesses in one of his more popular memes, "I had fun once. It was awful."

December 22, 2022

I hope you're hunkered down, dear bees, since the entire country is about to plunge into serious subzero temperatures, all the way down into Florida, even. The polar vortex is back. (Or what the Canadians call "arctic overflow.") Just in time for the holidays. Montana, Wyoming, and co. are already buried deep under white-out conditions. And this icy freight train should hit us on Christmas Eve. (Though sadly there is no actual snow forecast for New York City, so it likely won't be a white Christmas—just a fricken freezing one.) Thankfully, we have enough wine and fixins to ride out a claustrophobic week ahead. (With strong shades of the lockdown.)

January 1, 2023

Another year behind us. Another spin around the sun lies ahead.

After another Christmas with no tree. (Our third in a row.) Instead, we strung fairy lights around the wall of the living room, like a cheery kind of glowing mycelium. Somehow this feels more appropriate, in these more ecologically attuned times. (Albeit, less of an olfactory experience.) Christmas itself was bitterly cold, plunging to that territory on the temperature scale where Celsius and Fahrenheit begin to resemble each other—around minus 26 degrees. Not a comfortable place for a rendezvous, by anyone's reckoning. We didn't leave the apartment for several days and gorged on slow-cooked pork and homemade Christmas pudding. Given the bi-polar climate these days, however, New Year's Eve was relatively warm—around 6 degrees Celsius. (Even though I have been here for nearly two decades, I still think in metric units.) As a result of this leap above freezing, a thick fog hung over the city. Fairly Dickensian. In the Park, a blanket of mist covered the lake, where the warmer air met the frozen surface. Humidity was 99 percent, and drizzle did not so much as *fall* on all the tourists milling about in the gray soup, as simply appear, emerging from their garments, which simply transformed into dampness.

A couple of hours before midnight, heavy rain finally arrived. For a few seconds I watched footage of drenched tourists, penned like livestock in Times Square, but I had to turn it off. The whole spectacle was too sad, even if it's yet more evidence of an incremental return to normal. (This was the first year that no restrictions were placed on those foolhardy and unfathomable souls who that think ringing in the New Year compressed with tens of thousands of strangers in the middle of Manhattan, with barely a port-a-loo in sight, is a pilgrimage worth making.) Surprisingly, the municipal fireworks display went ahead, including, a local one, a few blocks away, near Strawberry Fields. (I suppose they now have the technology to protect fireworks from rain.) The deep booms were constant, rattling the windows, terrifying the city's pets, and amplified by the fog still muffling the island. Even those blessedly free of fresh memories of Ukraine, or more genetic memories of Napoleonic sea battles, could be forgiven for feeling anxious among such a battery.

All the more surprising, then, that the first day of 2023 was so clear, and blue, and tranquil. We walked through the Park, which was vibrant with animal life. (Apparently the local critters had already recovered from any trauma caused by the ruckus last night.) Hawks circled and soared, their belly feathers glowing gold in the morning sun. Squirrels foraged, and familiar birds were happy to brunch on the seeds we held out on our palms: titmice, nuthatches, chickadees, and even a cardinal, who perched for several minutes, taking his time to savor his meal. (Much to the chagrin of the smaller species.) We even spotted a baby racoon snoozing in a tree hollow.

By mid-afternoon, I was wearing a t-shirt on the communal deck of my building and recalling a twice-underlined passage in my trusty bee-centric bible: a moment from Maeterlinck that has since become something of a mantra for me: "It may be that these things are all vain." ("These things" being all the vital elements and experiences of this world.) Indeed, he continues, it may well be "that our own spiral of light, no less than that of the bees, has been kindled for no other purpose save that of amusing the darkness." When the weather is this unseasonal, I worry about the trees' becoming confused and then budding too early. But then I recall that life in any form, no matter how confused or disoriented, contains a seed of light that contains its own germinal purposes. Even if, as the man of letters so poignantly notes, it exists solely to entertain the void.

Basking in the winter sun, like a pale flower awaiting the industrious touch of the bee, I also can't deny it does wonders for the spirit to begin the year with a healthy dose of vitamin D.

Indeed, after a less-than-healthy lunch of Gruyère and stewed apple puff pastry—followed by some penance in the form of *qigong* and *xi* breathing—I feel ready to take on this new chapter. Of course, for someone raised in an "Irish" home—in the sense that no sense of hope or optimism was to be spoken out loud, lest the vindictive spirits decide to prove you wrong—it's almost taboo to express a positive outlook toward whatever comes. So, I won't get carried away. I won't make assumptions.

But I will say that right now—for this very moment—I feel *good*. And that in itself is something worth celebrating and cherishing. (Especially if I can keep carrying this feeling forward and share it with others.)

Yes, dear bees. In the spirit of the power of Now, I don't mind saying that right this second—at 5:09 P.M., on January 1, the year of their Lord, 2023 (even before fixing myself an *aperitivo*)—I feel good.

Good enough to do a little dance to a silly song.

Good enough to write an entire opera. My upbeat equivalent of Don Giovanni.

Good enough to wrestle a tiger. (In a playfight fashion since I don't want to hurt the poor creature!)

Good enough even to do a little dance . . . then write an opera . . . *and then* wrestle a tiger.

Printed in the USA
CPSIA information can be obtained
at www.ICGtesting.com
JSHW022234201124
73984JS00003B/18